Effective Coating Barriers for Protection of Reinforced Concrete

Effective Coating Barriers for Protection of Reinforced Concrete

Guest Editors

Junjie Wang
Hongwei Lin

MDPI

Basel • Beijing • Wuhan • Barcelona • Belgrade • Novi Sad • Cluj • Manchester

Guest Editors

Junjie Wang
Department of Civil Engineering
Tsinghua University
Beijing
China

Hongwei Lin
School of Civil Engineering
Beijing Jiaotong University
Beijing
China

Editorial Office
MDPI AG
Grosspeteranlage 5
4052 Basel, Switzerland

This is a reprint of the Special Issue, published open access by the journal *Coatings* (ISSN 2079-6412), freely accessible at: https://www.mdpi.com/journal/coatings/special_issues/concrete_coatings.

For citation purposes, cite each article independently as indicated on the article page online and as indicated below:

Lastname, A.A.; Lastname, B.B. Article Title. *Journal Name* **Year**, *Volume Number*, Page Range.

ISBN 978-3-7258-6732-5 (Hbk)
ISBN 978-3-7258-6733-2 (PDF)
https://doi.org/10.3390/books978-3-7258-6733-2

Contents

Preface

The imperative need to extend the service life of reinforced concrete in aggressive environments—epitomized by landmarks like the Hong Kong–Zhuhai–Macau Bridge—demands revolutionary approaches to corrosion protection. This reprint synthesizes cutting-edge research on effective coating barriers for concrete, with a dedicated focus on structures incorporating recycled materials and enduring marine exposure. As Guest Editors, we present this Special Issue to equip engineers and researchers with science-backed strategies for safeguarding century-scale infrastructure. The relentless challenge of ensuring the durability and longevity of reinforced concrete structures in aggressive environments has never been more pressing. As global infrastructure ages and sustainability demands rise, the integration of recycled materials in concrete—while environmentally prudent—introduces complex durability concerns. This reprint, "Effective Coating Barriers for Protection of Reinforced Concrete", confronts these issues head-on, focusing on cutting-edge coating technologies designed to shield concrete from corrosive threats, particularly in settings demanding exceptional service life.

Conventional coatings falter when protecting recycled-material concrete due to exacerbated porosity, chloride ingress, and carbonation. This Special Issue compiles pioneering research and practical insights aimed at engineers tasked with safeguarding infrastructures and researchers driving innovation in materials science. This collection addresses these vulnerabilities through three interconnected frontiers:

1. Advanced Coating Systems for Sustainable Concrete
   Smart nano SiO_2-modified hybrid coatings
   Slurry-coating treatments optimizing recycled aggregates' water resistance
   Bio-waste-derived carbon doping for multifunctional concrete
2. Corrosion Inhibition and Monitoring
   Novel perimidine derivatives inhibiting steel corrosion in concrete pore solutions
   Electrochemical evaluation frameworks for half-cell potential measurements
3. Cementitious Matrix Innovations
   Double-sulfate composites enhancing early strength
   Temperature-resilient gypsum–cement hybrids
   Diatomaceous seawater effects on blended cement hydration

This reprint responds to critical gaps in protecting sustainable infrastructure. By validating coatings' compatibility with recycled aggregates, introducing bio-derived modifiers, and redefining inhibitor chemistry, we provide actionable solutions for engineers designing marine megaprojects. Simultaneously, researchers gain fundamental insights into nano-modification mechanisms enhancing barrier longevity, corrosion inhibitor kinetics in alkaline environments, and hydration pathways under biological/seawater exposure.

The impetus for this reprint stems from an urgent convergence of factors: escalating environment-induced degradation, the proliferation of sustainable construction practices, and society's reliance on century-scale infrastructure. Traditional protection methods often fall short when concrete incorporates recycled constituents, which can accelerate chloride ingress or carbonation. Here, we present coatings not merely as barriers but as engineered systems that adapt to these nuances, ensuring resilience without compromising ecological goals.

We extend profound gratitude to the contributing authors—leaders in concrete, corrosion science, civil engineering, and chemical engineering—whose rigorous work forms the backbone of this volume. Their dedication to unraveling the synergies between coatings, concrete matrices, and environmental stressors has yielded actionable knowledge. We also acknowledge the invaluable efforts of peer reviewers, whose scrutiny elevated the quality of each submission, and our institutional partners for fostering interdisciplinary collaboration.

To engineers on the front lines of megaprojects and researchers pioneering tomorrow's solutions, may this reprint serve as both a compass and catalyst. As sea levels rise and infrastructure ages, the quest for unyielding concrete protection continues, and it is our collective responsibility to fortify the built world for generations to come.

Junjie Wang and Hongwei Lin
Guest Editors

coatings

Editorial

Effective Coating System Should Be Applied to Concrete with Recycled Waste Materials

Junjie Wang

Department of Civil Engineering, Tsinghua University, Beijing 100084, China; junjiewang@tsinghua.edu.cn

With global concerns over increasing CO_2 emissions, many countries have set up different strategies to achieve net zero CO_2 emissions. One important way to reduce CO_2 emissions is to use recycled waste materials, such as recycled aggregate concrete [1–14], recycled cement [15–17], and other recycled materials [18–24]. However, the physical properties of recycled materials might not be as good as the original ones [1–5]. For example, the recycled aggregates from demolished construction waste usually contain a lot of adhered mortar with a high porosity and micro cracks, which inevitably decrease the physical properties of concrete comprising 100% recycled aggregates, resulting in it being less durable under different environmental conditions (such as salt attack and carbonation). Thus, concrete that is primarily composed of recycled waste materials (e.g., >60%) is currently not allowed to be used in structural elements and can only be used in unimportant and non-structural elements.

As we know, concrete is one of the mostly widely used construction materials. Demands for raw materials such as natural stones, river sand, and limestone have been increasing due to needs for increased concrete production [16,19,22]. In local areas raw materials may be consumed at a rapid pace when there is a limited local supply. This represents an application for concrete that is primarily composed of recycled waste materials from demolition as well as other types of waste.

Steel-reinforced concrete elements may suffer steel corrosion during their service life. This corrosion is caused by the depassivation of steel by chloride ions in exposed environments [25–30] and concrete carbonation by the CO_2 in the atmosphere [31–33]. Other types of concrete deterioration include sulfate attack by the sulfate ions in water [34] and erosion caused by water. Additionally, the presence of micro cracks could accelerate the penetration of these harmful substances into concrete elements and could thereby accelerate the deterioration of concrete elements and decrease the service life of concrete structures [27,30,35]. When the composition of concrete most comprises recycled waste materials, concrete structures may have an even shorter service life, representing the main concern over using concrete with recycled waste materials.

Given that the mechanical properties have been fulfilled and the coating system can provide a long life and sustained lasting effects, in the author's opinion, through applying an effective coating system on the surface of concrete that is mostly composed of recycled waste materials, the deteriorated durability property of the concrete caused by recycled waste materials can be rectified.

An effective coating system [36–38] could provide protection to concrete made from recycled waste materials. Effective coating systems include protective coatings on concrete surfaces and steel surfaces and protection materials that can penetrate into the concrete. Coating materials include silane, corrosion inhibitors, epoxy, or other waterproof materials. Other types of innovative coatings are also encouraged. When silane is brushed or sprayed on a concrete surface, it gradually penetrates into the concrete inside and reacts with the water inside the concrete in an alkaline environment, resulting in highly reactive hydroxylases being formed that will react with the hydroxyl groups, forming a water-repellent molecular layer [39–41]. For example, Basheer et al. [42] showed that silane impregnation could extend the service life of concrete under freeze–thawing cycles by more than 100%.

However, there is still limited understanding on the long-term degradation of coating materials and the long-term interaction mechanisms between these materials and waste-based concrete, especially the long-term degradation mechanisms [43,44] of these materials on the surface and inside waste-based concrete. In addition, de-bonding between coating materials and concrete or steel surfaces could happen during the expected service life of waste-based concrete.

Future research on the following subjects is encouraged:

- Long-life performance and deterioration mechanisms of coating systems on the surfaces of waste-based concrete in different exposure environments;
- Long-term adhesion and compatibility between coating systems and waste-based concrete;
- Multiple coating systems and their interactions with waste-based concrete.

Funding: The author wish to acknowledge the financial support provided by the National Natural Science Foundation of China (Nos. 52008232, 52038004).

Institutional Review Board Statement: Not applicable.

Informed Consent Statement: Not applicable.

Conflicts of Interest: The author declares no conflict of interest.

References

1. Xie, J.; Huang, L.; Guo, Y.; Li, Z.; Fang, C.; Li, L.; Wang, J. Experimental study on the compressive and flexural behaviour of recycled aggregate concrete modified with silica fume and fibres. *Constr. Build. Mater.* **2018**, *178*, 612–623. [CrossRef]
2. Wang, H.; Sun, X.; Wang, J.; Monteiro, P.J.M. Permeability of concrete with recycled concrete aggregate and pozzolanic materials under stress. *Materials* **2016**, *9*, 252. [CrossRef] [PubMed]
3. Wang, H.; Wang, J.; Sun, X.; Jin, W. Improving performance of recycled aggregate concrete with superfine pozzolanic powders. *J. Cent. South Univ.* **2013**, *20*, 3715–3722. [CrossRef]
4. Wang, J.; Liu, E. Additions of different calcium carbonate minerals in cement to increase material greenness. In Proceedings of the 1st International Conference on Innovation in Low-Carbon Cement & Concrete Technology ILCCC2019, London, UK, 24–26 June 2019.
5. Xie, J.; Zhao, J.; Wang, J.; Fang, C.; Yuan, B.; Wu, Y. Impact behaviour of fly ash and slag-based geopolymeric concrete: The effects of recycled aggregate content, water-binder ratio and curing age. *Constr. Build. Mater.* **2022**, *331*, 127359. [CrossRef]
6. Xie, J.; Zhao, J.; Wang, J.; Huang, P.; Liu, J. Investigation of the high-temperature resistance of sludge ceramsite concrete with recycled fine aggregates and GGBS and its application in hollow blocks. *J. Build. Eng.* **2021**, *34*, 101954. [CrossRef]
7. Wang, J.; Liu, E. Rheological and calorimetric behaviours of cement paste with SiO_2 based and $CaCO_3$ based mineral additions. In Proceedings of the 2nd International Conference on UHPC Materials and Structures UHPC 2018, Fuzhou, China, 7–10 November 2018; pp. 200–210.
8. Wang, H.; Wang, J.; Sun, X.; Chen, J. Chloride diffusion characteristics of the new interface transition zone in recycled aggregate concrete. *Adv. Mater. Res.* **2011**, *261–263*, 104–110. [CrossRef]
9. Wang, J. Micro-Mechanism of Improving Mechanical Property and Durability of Recycled Aggregate Concrete Based on Material Meso-Structures and Micro-Structures. Master Thesis, Zhejiang University, Hangzhou, China, 2011; p. 81. (In Chinese).
10. Wang, J.; Xie, J.; Wang, C.; Fang, C.; Liu, F. Study on the optimum initial curing condition for fly ash and GGBS based geopolymer recycled aggregate concrete. *Constr. Build. Mater.* **2020**, *247*, 118540. [CrossRef]
11. Wang, J.; Xie, J.; He, J.; Sun, M.; Zhao, J. Combined use of silica fume and steel fibre to improve fracture properties of recycled aggregate concrete exposed to elevated temperature. *J. Mater. Cycles Waste Manag.* **2020**, *22*, 862–877. [CrossRef]
12. Xie, J.; Wang, J.; Rao, R.; Wang, C.; Fang, C. Effects of combined usage of GGBS and fly ash on workability and mechanical properties of alkali activated geopolymer concrete with recycled aggregate. *Compos. Part B Eng.* **2019**, *164*, 179–190. [CrossRef]
13. Xie, J.; Chen, W.; Wang, J.; Fang, C.; Zhang, B.; Liu, F. Coupling effects of recycled aggregate and GGBS/metakaolin on physicochemical properties of geopolymer concrete. *Constr. Build. Mater.* **2019**, *226*, 345–359. [CrossRef]
14. Xie, J.; Wang, J.; Zhang, B.; Fang, C.; Li, L. Physicochemical properties of alkali activated GGBS and fly ash geopolymeric recycled concrete. *Constr. Build. Mater.* **2019**, *204*, 384–398. [CrossRef]
15. Xu, L.; Wang, J.; Li, K.; Lin, S.; Li, M.; Hao, T.; Ling, Z.; Xiang, D.; Wang, T. A systematic review of factors affecting properties of thermal-activated recycled cement. *Resour. Conserv. Recycl.* **2022**, *185*, 106432. [CrossRef]
16. He, Z.; Zhu, X.; Wang, J.; Mu, M.; Wang, Y. Comparison of CO_2 emissions from OPC and recycled cement production. *Constr. Build. Mater.* **2019**, *211*, 965–973. [CrossRef]

17. Wang, J.; Mu, M.; Liu, Y. Recycled cement. *Constr. Build. Mater.* **2018**, *190*, 1124–1132. [CrossRef]
18. Wang, Y.; He, H.; Wang, J.; Lia, F.; Ding, Y.; Xu, L. Effect of aggregate micro fines in machine-made sand on bleeding, autogenous shrinkage and plastic shrinkage cracking of concrete. *Mater. Struct.* **2022**, *55*, 106. [CrossRef]
19. Wang, J.; Liu, E. Upcycling waste seashells with cement: Rheology and early-age properties of Portland cement paste. *Resour. Conserv. Recycl.* **2020**, *155*, 104680. [CrossRef]
20. Wang, Y.; Tian, Y.; Wang, J. Effects of carbide slag and CO_2 curing on physical properties of gypsum plaster. *ACI Mater. J.* **2020**, *117*, 169–178. [CrossRef]
21. He, H.; Wang, Y.; Wang, J. Compactness and hardened properties of machine-made sand mortar with aggregate micro fines. *Constr. Build. Mater.* **2020**, *250*, 118828. [CrossRef]
22. Wang, J.; Liu, E.; Li, L. Characterization on the recycling of waste seashells with Portland cement towards sustainable cementitious materials. *J. Clean. Prod.* **2019**, *220*, 235–252. [CrossRef]
23. Xie, J.; Liu, J.; Liu, F.; Wang, J.; Huang, P. Investigation of a new lightweight green concrete containing sludge ceramsite and recycled fine aggregates. *J. Clean. Prod.* **2019**, *235*, 1240–1254. [CrossRef]
24. Hang, H.; Wang, Y.; Wang, J. Effects of aggregate micro fines (AMF), aluminum sulfate and polypropylene fiber (PPF) on properties of machine-made sand concrete. *Appl. Sci.* **2019**, *9*, 2250.
25. Xie, J.; Wang, J.; Li, M.; Xu, L.; Xiang, D.; Wang, Y.; He, H.; Zhu, Y.; Zhao, J. Estimation of chloride diffusion coefficient from water permeability test of cementitious materials. *Constr. Build. Mater.* **2022**, *340*, 127816. [CrossRef]
26. Wang, J.; Xie, J.; Wang, Y.; Liu, Y.; Ding, Y. Rheological properties, compressive strength, hydration products and microstructure of seawater-mixed cement pastes. *Cem. Concr. Compos.* **2020**, *114*, 103770. [CrossRef]
27. Xie, J.; Wang, J.; Liu, Y.; Wang, Y. Comparison of three different methods for measuring chloride transport in predamaged concretes. *J. Mater. Civ. Eng.* **2020**, *32*, 04020033. [CrossRef]
28. Liu, J.W.E. The relationship between steady-state chloride diffusion and migration coefficients in cementitious materials. *Mag. Concr. Res.* **2020**, *72*, 1016–1026.
29. Wang, J.; Liu, E.; Li, L. Multiscale investigations on hydration mechanisms in seawater OPC paste. *Constr. Build. Mater.* **2018**, *191*, 891–903. [CrossRef]
30. Wang, J. Steady-state chloride diffusion coefficient and chloride migration coefficient of cracks in concrete. *J. Mater. Civ. Eng.* **2017**, *29*, 0417117. [CrossRef]
31. Wang, J.; Xu, L.; Li, M.; He, H.; Wang, Y.; Xiang, D.; Lin, S.; Zhong, Y.; Zhao, H. Effect of pre-carbonation on the properties of cement paste subjected to high temperatures. *J. Build. Eng.* **2022**, *51*, 104337. [CrossRef]
32. Wang, Y.; He, F.; Wang, J.; Wang, C.; Xiong, Z. Effects of calcium bicarbonate on the properties of ordinary Portland cement paste. *Constr. Build. Mater.* **2019**, *225*, 591–600. [CrossRef]
33. Wang, Y.; He, F.; Wang, J.; Hu, Q. Comparison of effects of sodium bicarbonate and sodium carbonate on the hydration and properties of Portland cement paste. *Materials* **2019**, *12*, 1033. [CrossRef]
34. Xie, J.; Wang, J.; Wang, C.; Fang, C.; Li, L.; He, J. Sulfate resistance of recycled aggregate concrete with GGBS and fly ash-based geopolymer. *Materials* **2019**, *12*, 1247. [CrossRef] [PubMed]
35. Wang, J.; Basheer, P.A.M.; Nanukuttan, S.V.; Long, A.E.; Bai, Y. Influence of service loading and the resulting micro-cracks on chloride resistance of concrete. *Constr. Build. Mater.* **2016**, *108*, 56–66. [CrossRef]
36. Mu, M.; Ou, C.; Wang, J.; Liu, Y. Surface modification of prototypes in fused filament fabrication using chemical vapour smoothing. *Addit. Manuf.* **2020**, *31*, 100972. [CrossRef]
37. Dang, Y.; Ning, X.; Kessel, A.; McVey, E.; Pace, A.; Shi, X. Accelerated laboratory evaluation of surface treatments for protecting concrete bridge decks from salt scaling. *Constr. Build. Mater.* **2014**, *55*, 128–135. [CrossRef]
38. Pan, X.; Shi, Z.; Shi, C.; Ling, T.C.; Li, N. A review on concrete surface treatment Part I: Types and mechanisms. *Constr. Build. Mater.* **2017**, *132*, 578–590. [CrossRef]
39. Du, Z. *Organosilicon Chemistry*; Higher Education Press: Beijing, China, 1990; pp. 184–206. (In Chinese)
40. American Concrete Institute. *Guide for the Use of Polymers in Concrete*; ACI: Farmington Hills, MI, USA, 2009.
41. Li, K.; Jing, W.; Yang, R. Review on silane impregnation of concrete surface and itslong-term hydrophobic performance. *J. Chin. Ceram. Soc.* **2019**, *47*, 10.
42. Basheer, L.; Cleland, D.J. Freeze–thaw resistance of concretes treated with pore liners. *Constr. Build. Mater.* **2006**, *20*, 990–998. [CrossRef]
43. Wang, Y.; Lu, H.; Wang, J.; He, F. Effects of highly crystalized nano C-S-H particles on performances of Portland cement paste and its mechanism. *Crystals* **2020**, *10*, 816. [CrossRef]
44. Wang, Y.; Yu, J.; Wang, J.; Guan, X. Effects of aluminum sulfate and quicklime/fluorgypsum ratio on the properties of calcium sulfoaluminate (CSA) cement based double liquid grouting materials. *Materials* **2019**, *12*, 1222. [CrossRef]

coatings

MDPI

Article

Physicomechanical Properties of Gypsum with Mineral Additions at Elevated Temperatures

Junjie Wang [1] and Engui Liu [2,*]

1 Department of Civil Engineering, Tsinghua University, Beijing 100084, China; junjiewang@tsinghua.edu.cn
2 Department of Civil Engineering, Xi'an Jiaotong-Liverpool University, Suzhou 215123, China
* Correspondence: engui.liu@xjtlu.edu.cn

Abstract: Gypsum, from either nature or industrial by-products, can be a lower-cost and cleaner alternative binder to Portland cement used in construction projects, such as affordable housing in developing countries. Although various building products have successfully used gypsum as the binder, some drawbacks of this material have still been claimed, for example, in the aspects of mechanical strength and some other physical properties. Using mineral additions to gypsum seems to be a possible solution to create composite gypsum with improved properties. This work has investigated the possibility of two common minerals (silica flour and talc powder) in modifying composite gypsum's physical and mechanical performance at elevated temperatures (100–1100 °C), including hydration, strength, thermal conduction and stability, and microstructure. The results suggest that 10% gypsum replacement by silica flour or talc powder modifies gypsum's physical and mechanical properties, with silica flour performing better than talc powder. The performance of composite gypsum at elevated temperatures depends on the treatment temperature and reflects the combined effects of gypsum phase change and the filler effects of silica flour or talc powder. Thermal treatment at $\leq$200 °C increased the thermal resistance of all gypsum boards but decreased their compressive strength. Thermal treatment at $\geq$300 °C significantly increased the compressive strength of gypsum with silica flour and talc powder but induced intensive microcracks and thus failed the thermal insulation. This investigation indicates that silica flour can potentially raise the mechanical performance of gypsum. At the same time, talc powder can hold water and lubricate, which may help with the continuous hydration of gypsum phases and the rheology of its mixtures.

Keywords: gypsum; mineral additions; fillers; silica; talc; thermal treatment

1. Introduction

Portland cement has been the leading cementitious material for construction worldwide. It will likely dominate for the following decades, given that it is complicated, if not impossible, to replace Portland cement with alternative cement entirely [1,2]. Still, considering the many environmental impacts of using Portland cement and its relatively high cost in developing countries, other cementitious materials with lower prices and easier local availability may increasingly be used in construction, such as for low-cost houses [3]. Gypsum can be a cementitious material as it can be sourced from nature in many countries or from various industries where gypsum is produced as solid waste [4]. Other local materials, e.g., silicate and clay minerals, can also be combined with gypsum as fillers to prepare binders with improved properties [5–7].

Gypsum has been successfully used in multiple construction fields, such as cement production, interior panels and decorations, and self-leveling composites. Gypsum boards also have fire-resistance capacity owing to their phase change from gypsum dihydrate ($CaSO_4{\cdot}2H_2O$) to hemihydrate ($CaSO_4{\cdot}0.5H_2O$) and anhydrite ($CaSO_4$) when exposed to fire or intense heat [8]. These states could be interchangeable depending on the thermal exposure and availability of moisture. The phase change process usually comes with water vapor release and pore structure change, leading to decreased heat conduction and possibly enhanced strength in the gypsum building elements, extending their protection time before structural collapse during the fire. In addition, gypsum is a renewable resource that can be obtained through recycling construction waste, among other methods. In addition, gypsum has relatively low raw material costs and production costs compared to cement. Furthermore, the manufacturing process of gypsum requires less energy than that of cement, which helps reduce the carbon footprint and environmental impact.

However, the properties of gypsum may be further improved in aspects such as the processing time of fresh mixtures [9], mechanical strength [10], and thermal stability after exposure at elevated temperatures [11], making it more suitable for use in housing projects. Bicer and Kar [12] used expanded polystyrene and tragacanth within gypsum mixtures, finding that the thermal conductivity and compressive strength decreased simultaneously. Gencel et al. [13] reported an effective way to improve mechanical and thermal resistance by adding diatomite and polypropylene fibers. Camarini et al. [14] and Lanzón and García-Ruiz [15] added citric acid to the gypsum plaster to improve its workability and physical properties. Khalil et al. [16] reported that adding natural sand, silica fume, and anhydrous silica gel could improve gypsum plaster composites' physical and mechanical properties. They also indicated that some waste additives, such as unburnt rice husk, blast furnace slag, calcium carbonate, and polyvinyl alcohol polymer, could improve gypsum's mechanical properties. Pundir et al. [17] reported that a sulfonated melamine formaldehyde condensate-based superplasticizer could accelerate the hydration of gypsum and form a dense and compacted structure of gypsum plaster. Tokarev et al. [18] added single-wall carbon nanotubes with Portland cement, microsilica, and metakaolin to the gypsum plaster. They reported that the microstructure and mechanical properties were improved in the gypsum. Zhu et al. [19] reported a significant improvement in the mechanical properties of gypsum-based composites with the addition of polyvinyl alcohol and polypropylene fibers. Sophia and Sakthieswaran [6] showed that gypsum plaster with enhanced water and thermal stability could be produced through the addition of zeolite (natural and calcined) and shell powders (egg, conch, cuttlebone, and scallop).

This work investigated the possibility of adding two regular minerals, silica flour and talc powder, to gypsum pastes, intending to improve gypsum's physical and mechanical performances. These two types of minerals are readily available in many parts of the world and have been generally used in cementitious material fields as mineral fillers [20–23]. Silica flour can also be equivalently obtained in the manufactured sand industry as a by-product (sand dust), the fine residue from crushing rocks into manufactured sand for construction use [24]. As a hydrated magnesium silicate clay mineral, talc powder demonstrated the potential of being incorporated into cementitious compositions owing to its properties such as water absorption [25,26] and lubrication [27] that may help with the hydration and rheology of cementitious systems. This study was focused on the physical and mechanical aspects of blended gypsum paste with silica flour and talc powder additions. It also looked at gypsum property changes after being treated at elevated temperatures. The properties investigated included compressive strength, thermal conductivity, weight loss, phase assemblage, and the composite gypsum paste microstructure to prove the concept.

2. Experimental Program

Raw materials: Gypsum powder, talc powder, and silica flour were used as received from local suppliers. The samples were dried in a vacuum drying oven at 40 °C for 48 h, followed by sieving through 200 mesh screens to obtain the specimens intended for XRD and XRF analysis. Table 1 shows their rough chemical compositions measured by an X-ray fluorescence (XRF) spectrometer (Rigaku NEC GC). Gypsum has the oxides CaO and SO_3, while talc includes CaO, SiO_2, and MgO, and silica flour has SiO_2 as the main compositional component. Figure 1 shows that gypsum powder has the smallest particle size on average, followed by talc powder and silica flour. Figure 2 plots the XRD patterns of the three minerals, indicating that gypsum powder mainly contains hemihydrate ($CaSO_4{\cdot}0.5H_2O$)

and some calcite, while talc powder has talc and dolomite together with some quartz, and, lastly, silica flour contains quartz entirely. The high LOI in gypsum and talc powder could be due to the existence of moisture absorbed from the air.

Table 1. Chemical compositions of gypsum powder, talc powder, and silica flour, indicated by X-ray fluorescence spectroscopy in mass %.

Oxides	Gypsum Powder	Talc Powder	Silica Flour
CaO	32.60	21.40	0.24
SiO_2	0.68	17.40	94.00
SO_3	35.30	0.10	0.05
SrO	0.24	—	—
ZrO_2	0.56	0.42	0.34
MnO	—	0.07	—
MgO	—	12.00	0.13
Fe_2O_3	0.09	1.27	0.05
LOI	30.53	47.34	5.19

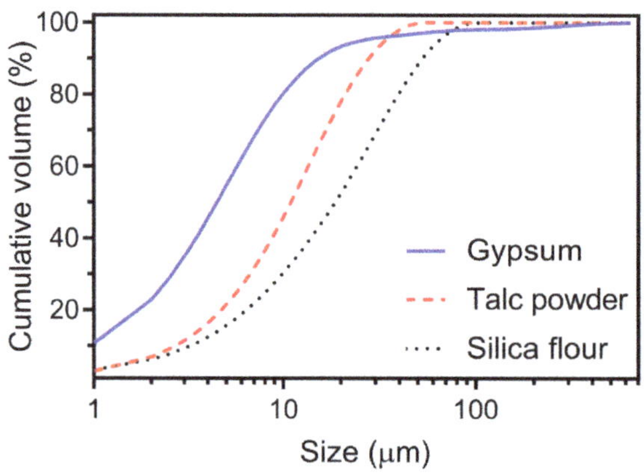

Figure 1. Particle size distribution of gypsum powder, talc powder, and silica flour.

Gypsum paste preparation: Three types of gypsum pastes were prepared in this work. The first one, made with pure gypsum, was used as the reference (named GYP). TAC and SIL were blended gypsum pastes made with 90 wt% gypsum and 10 wt% talc powder or silica flour, respectively. The water content was constant at 60% by powder mass in the paste. To prepare the mixture, gypsum powder and talc powder (or silica flour) were dry-mixed well for 1 min in a lab mixer at a medium mixing speed. After that, deionized water was added to the dry blend and mixed with high shear for 2 min until completion. The fresh paste was cast immediately into cube molds (25 mm × 25 mm × 25 mm) and slab molds (300 mm × 300 mm × 25 mm) and then vibrated for 10 s to remove entrained air bubbles. The samples' curing was conducted in an environmental chamber at 20 °C and 97% R.H. The specimens were de-molded after 24 h of curing. Three samples were prepared for each group for the measurement of physical properties.

Thermal treatment: The samples were heated at a rate of 10 °C/min from room temperature to a target temperature. The target temperature was then held for 6 h before cooling back to room temperature. The target temperatures selected in this study were 100 °C, 200 °C, 300 °C, 600 °C, and 1100 °C. The samples kept at room temperature (about 25 °C) were used as references.

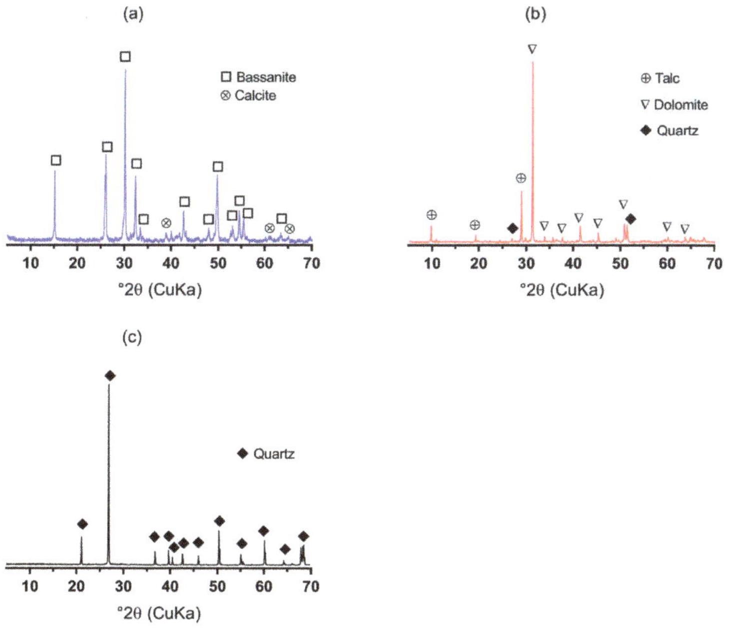

Figure 2. XRD pattern of (**a**) gypsum powder, (**b**) talc powder, and (**c**) silica flour.

Hydration heat: The hydration heat flow rate and total heat of hydration of the gypsum pastes were recorded by a standard isothermal calorimeter (Calmetrix I-Cal 8000). A total of 20 g of powder (pure gypsum, gypsum with silica flour, or gypsum with talc powder) and 12 g of deionized water were mixed manually for 30 s in the measurement cell. After that, the cell was immediately sealed and loaded into the calorimeter chamber. The hydration heat was monitored for 72 h.

Thermal conductivity: A standard heat flow meter in the field (TA instruments NETZSCH HFM 436/3/1 E Lambda) was utilized to measure the thermal conductivity of the slab samples (300 mm × 300 mm × 25 mm) after being treated at elevated temperatures. During the measurement, the temperature of the testing chamber was maintained at 25 °C.

Compressive strength: A universal testing machine was used to compress cube specimens (25 mm × 25 mm × 25 mm) for their strengths. The compression was operated in a displacement control mode with a loading rate of 0.02 mm/s. The compressive strength was based on the average of three values.

Thermogravimetric analysis (TGA): The TGA tests were conducted using a simultaneous thermogravimetric analyzer (Hitachi STA7300). About 10 mg of the grounded powder sample was loaded into a platinum pan, with another empty platinum pan used as the reference in the machine furnace [28,29]. The furnace was stabilized for 30 min to reach the isothermal condition (at 25 °C), followed by rising its temperature to 900 °C at a 10 °C/min rate. The nitrogen gas flow was the protection medium in the furnace during the testing of samples.

X-ray diffraction (XRD) and scanning electron microscopy (SEM): The selected gypsum paste sample (around 2 g) was manually ground in a laboratory mortar and pestle set with about 1 g of fine powder for XRD analysis. A standard diffractometer using Kα radiation was used to conduct the analysis. The sample underwent 10 scans in each measurement with the 2θ from 5° to 70° (PIXcel detector), which were performed using a PANalytical Empyrean diffractometer and operating at 45 kV and 40 mA in a Bragg–Brentano geometry. The SEM (Quanta 450 FEG) tests were conducted on selected sample pieces of gypsum paste. After being coated with gold for 240 s, each specimen was observed in the microscope chamber with a high vacuum and secondary electron mode for micromorphology.

3. Experimental Results

3.1. Hydration Heat

Figure 3 shows the heat flow rate and total released heat of hydration of the three gypsum mixtures (GYP, TAC, and SIL). The heat flow rate within and after 2 h is presented separately in Figure 3a,b to clearly show their different trends. Figure 3a shows that the first peak at 2–3 min indicates the initiation of gypsum dissolution in water. The second peak demonstrates that gypsum hydrates with water, which generates lots of heat [30,31]. Both peak heights were reduced when the gypsum was replaced by talc powder or silica flour compared to the pure gypsum mixture (GYP). The peak appearance time was also delayed in the TAC and SIL mixtures. The peak appeared 2 min and 5 min later in the SIL and TAC mixtures, respectively, than in the GYP mixture. The peak height value of the SIL mixture was higher than that of the TAC mixture, indicating that silica flour performed better on gypsum hydration than talc powder.

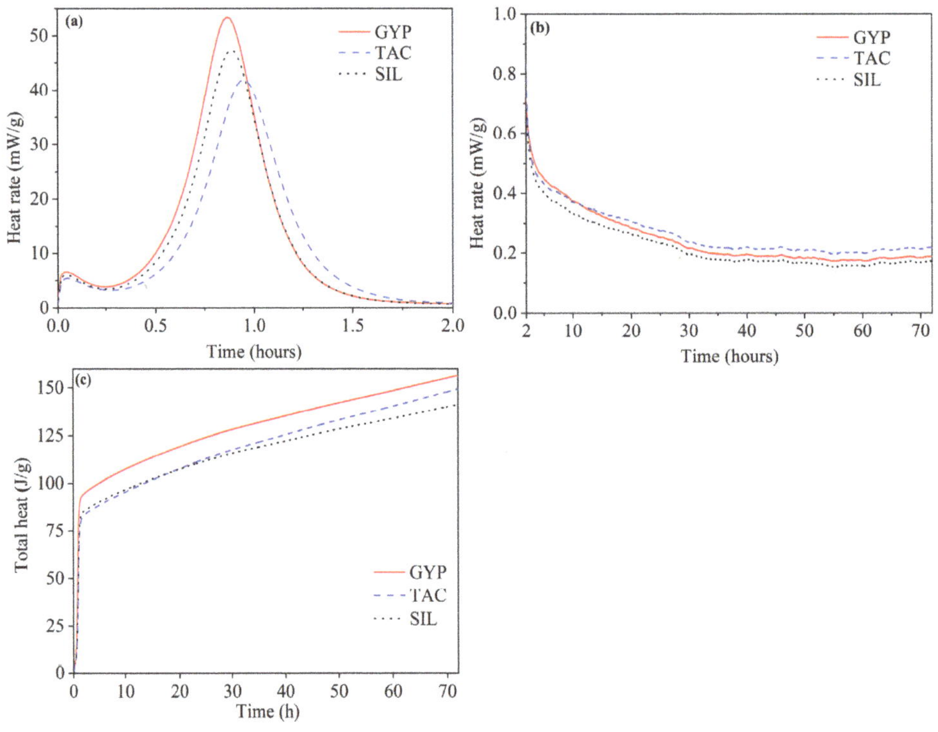

Figure 3. Hydration heat development of the three gypsum mixtures: (**a**) heat flow rate during the first 2 h, (**b**) heat flow rate from 2 to 70 h, and (**c**) total heat of hydration during the 72 h of measurement.

After about 10 h, as shown in Figure 3b, the TAC mixture exhibited the highest heat flow rate among the three (even higher than the pure gypsum mixture, GYP). The delayed hydration was the main reason for this phenomenon, and there was relatively more gypsum to hydrate after 10 h than in the other two mixtures. Figure 3c demonstrates the total related heat from the three mixtures during the 72 h of hydration. The TAC mixture started to generate more hydration heat than the SIL mixture at around the 25 h mark, continuing until the 72 h mark. This finding might reveal the role of talc powder in promoting the later-stage hydration of gypsum, possibly due to the release of absorbed water from talc that supported the continuous hydration of gypsum, similar to the well-known concept of internal curing in the field [32].

3.2. Compressive Strength

Figure 4 shows the compressive strength of gypsum pastes at elevated temperatures, with the standard temperature (25 °C) as the reference. The results at 1100 °C were unavailable because significant macrocracking formed in the specimen after the treatment. It can be seen that, at 25 °C, SIL samples had the highest strength among the three, more than twice that of GYP. The TAC samples had slightly lower compressive strength than the GYP samples. This result suggested that adding silica flour was influential in developing high-strength gypsum mixtures, possibly due to the higher hardness of silica than gypsum. Also, the average particle size of silica flour was bigger than that of gypsum, as shown in Figure 1. Thus, silica flour particles performed as a solid skeleton that increased the strength of the gypsum matrix [33]. Talc powder, however, was softer than gypsum, and, more importantly, it had physically adsorbed water on the particle surface due to its water-absorbing nature. Therefore, the interfacial region of talc and gypsum negatively influenced the strength of gypsum from this perspective.

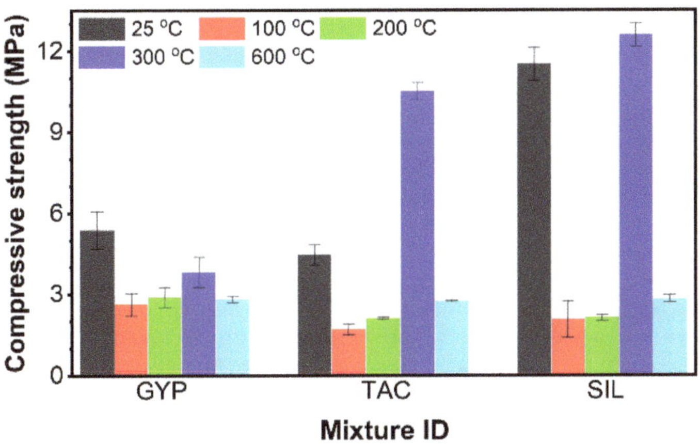

Figure 4. Compressive strength of hardened gypsum pastes at elevated temperatures.

With the increase in treatment temperature, the strength of all specimens decreased at 100 °C and 200 °C but increased remarkably at 300 °C and then decreased again at 600 °C. There was tremendous pore water loss, and gypsum dehydration happened at 100 °C and 200 °C, leading to a porous structure that weakened the strength of the gypsum matrix. After the 300 °C treatment, the dehydration of gypsum was likely completed, forming a more robust anhydrate phase that increased the strength of the gypsum matrix [11]. Also, talc particles lost their physically adsorbed water after the temperature treatment, turning into surface-dry fillers better embedded into the gypsum matrix [25]. Silica flour worked similarly to talc powder but contributed more to the strength of gypsum owing to its higher hardness. Furthermore, Figure 5 shows the microcracking (highlighted in blue) on the gypsum specimens after the 300 °C treatment, which might have contributed to energy and stress dissipation, particularly considering the strength of the gypsum matrix increased after the 300 °C treatment. The 600 °C treatment, however, developed detrimental microcracks in the specimens and thus significantly decreased their compressive strength values.

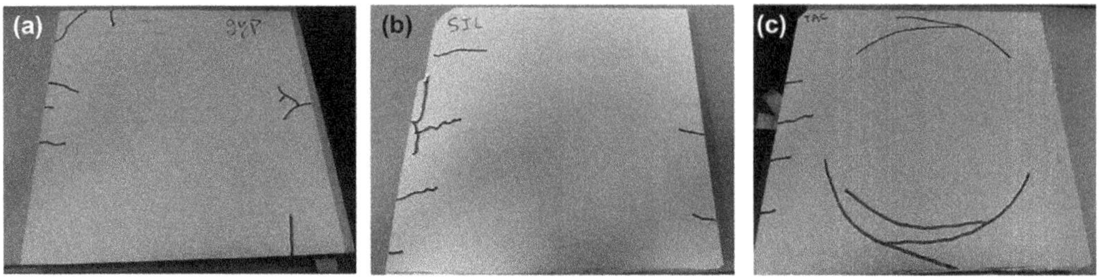

Figure 5. Microcracking in gypsum specimens after the 300 °C treatment: (**a**) GYP, (**b**) SIL, and (**c**) TAC.

3.3. Thermal Conductivity, Weight Loss, and Shrinkage

Figure 6a shows the thermal conductivity of gypsum specimens after being treated at 100 °C and 200 °C (25 °C as the reference). The specimens treated at 300 °C were not measurable due to the thermal cracking that affected the heat flow testing. At 25 °C (without heat treatment), the thermal conductivity of GYP specimens was the highest among the three, followed by the TAC and SIL specimens. The reason could be the more homogeneous structure and composition in the GYP matrix, which encouraged heat conduction. Talc powder's water adsorption was responsible for the higher thermal conductivity of TAC specimens than SIL. After the 100 °C treatment, all samples demonstrated reductions in thermal conductivity because the gypsum's dehydration left a porous structure in the matrix [34]. The GYP and TAC specimens performed similarly in terms of thermal conductivity, indicating that the material composition difference was less influential than the structural change. In contrast, the SIL specimens still had the lowest thermal conductivity, possibly owing to less water retained in the matrix than the GYP and TAC specimens. After the 200 °C treatment, all specimens reduced their thermal conductivity and demonstrated comparable performance regardless of their different material compositions. This phenomenon was due to the profound dehydration and water evaporation that left a more porous structure in the gypsum matrix, which played a leading role.

Figure 6b shows the weight loss of the gypsum specimens after treatment at elevated temperatures. Most weight losses happened before 300 °C, e.g., 15% at 200 °C and 20% at 300 °C, corresponding roughly to gypsum dihydrate becoming hemihydrate and then anhydrite. The weight loss amount from SIL specimens was the smallest among the three at all exposure temperatures, followed by the GYP and TAC specimens. The water absorption of talc powder was superior to that of silica flour in the gypsum mixtures and thus contributed to a much higher weight loss in the TAC specimens. Figure 6c shows the linear shrinkage of gypsum bar specimens after treatment at 100 °C and 200 °C. The shrinkage of the 300 °C treatment was unavailable because all the bar specimens broke after the heat treatment. At 100 °C, the GYP and SIL samples demonstrated a sensible shrinkage; however, the TAC specimens had no shrinkage but a slight expansion. This was possibly due to the rehydration of gypsum hemihydrate/anhydrite with the water harvested from the talc powder parallel to the dehydration [35]. At 200 °C, the TAC specimens shrank the most (due to a much more substantial water loss beyond gypsum dehydration) compared to the GYP and SIL specimens.

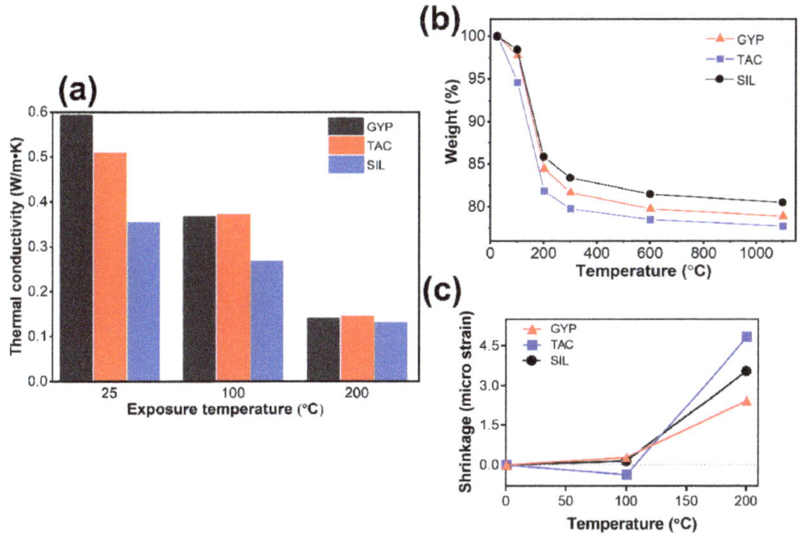

Figure 6. Thermal conductivity (**a**), weight loss (**b**), and linear shrinkage (**c**) of gypsum specimens after being treated at elevated temperatures.

3.4. TGA and XRD

TGA and XRD analyzed gypsum specimens after treatment at elevated temperatures to identify weight loss and phase information, as shown in Figures 7 and 8. In Figure 7, the weight loss in the 100–200 °C region was mainly due to matrix-confined water loss and gypsum dehydration [36], and in the 600–700 °C region, to the decomposition of carbonate and possibly talc [37]. The results in Figure 7a indicate that there was roughly about 20% water weight confined in the material matrix and gypsum dihydrate. After the 200 °C treatment, this amount of water was reduced by about 15%, as shown in Figure 7b, suggesting that almost all the gypsum dihydrate turned into hemihydrate. After the 300 °C treatment, the weight loss was further reduced in the specimens compared to the 200 °C treatment, as shown in Figure 7c, suggesting that the gypsum hemihydrate was dehydrated into anhydrite. The findings in TGA are also supported by the XRD results shown in Figure 8.

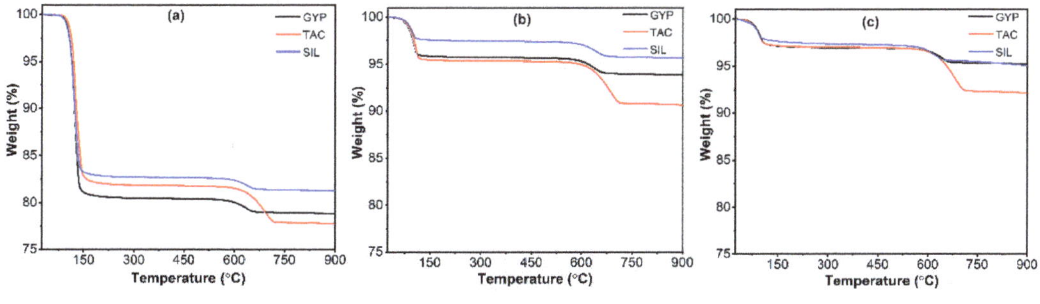

Figure 7. TGA results of gypsum specimens after being treated at elevated temperatures: (**a**) 25 °C as the reference, (**b**) 200 °C, and (**c**) 300 °C.

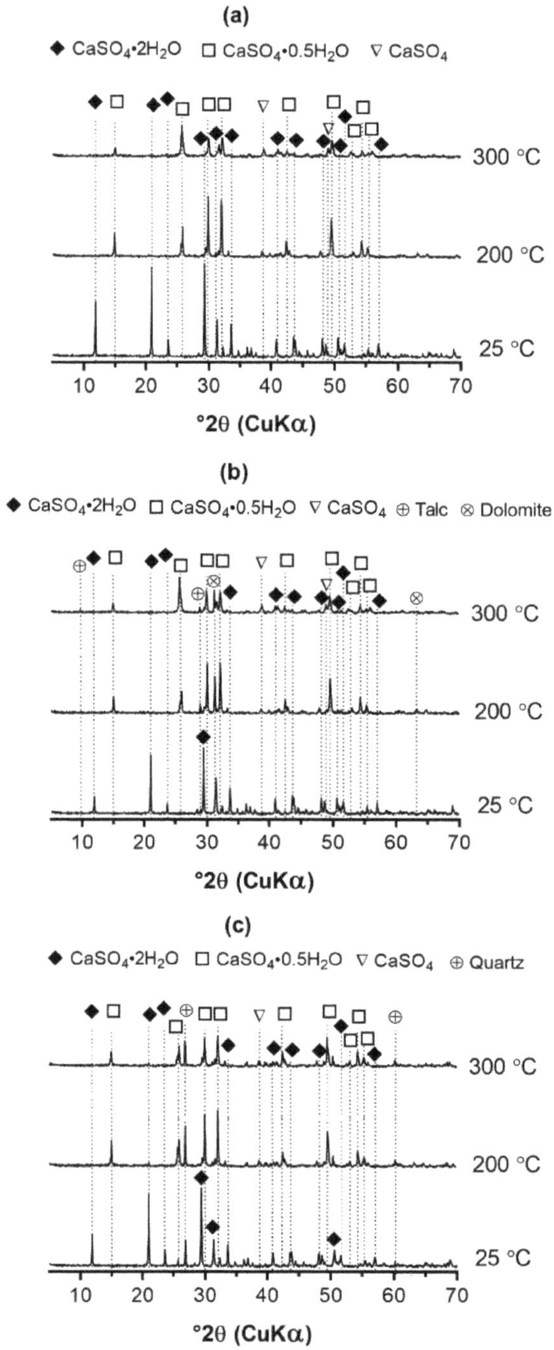

Figure 8. XRD patterns of gypsum specimens after being treated at 200 °C and 300 °C, with 25 °C as the reference: (**a**) GYP, (**b**) TAC, and (**c**) SIL.

The GYP specimen had the most weight loss at 25 °C (without thermal treatment), followed by the TAC and SIL specimens, as shown in Figure 7a. This weight loss was attributable to gypsum dehydration and matrix-confined water. The GYP specimen contained the most gypsum hydrate, demonstrating the most weight loss among the three. In the SIL and TAC specimens, there was 10% of gypsum replaced by silica flour or talc powder, and thus the amount of gypsum hydrate was reduced (resulting in a decreased weight loss). The free water in the matrix likely evaporated during the curing of the specimens. The amount of matrix-confined water in the TAC specimen was higher than in the SIL specimen, as talc powder adsorbed water better than silica flour due to its smaller average particle size and stronger water affinity. After the thermal treatment at 200 °C and 300 °C (Figure 7b,c), the weight loss in the SIL specimen was still the lowest among the three, while the weight loss values in the TAC and GYP specimens were comparable. The reason might be the water confinement by talc during the dehydration of gypsum, so this amount of water contributed to the weight loss after the thermal treatment at 200 °C and 300 °C in the TAC specimen.

It can be seen from Figure 8 that gypsum dihydrate was the mineral in the hydrated gypsum matrix without thermal treatment (25 °C). After the thermal treatment at 200 °C, the gypsum dihydrate lost one and a half moles of water and transformed into gypsum hemihydrate (with half a mole of water left). After the treatment at 300 °C, gypsum dihydrate was directly transformed into anhydrite ($CaSO_4$). Some anhydrite was further transformed into gypsum hemihydrate upon contact with vapor in the air during the cooling stage [38]. Talc and dolomite were shown in the TAC samples, and quartz was identified in the SIL specimens. There were no noticeable changes in their peaks with the thermal treatments.

3.5. Microstructural Morphology

This work's three types of mixtures demonstrated good mechanical strength at standard temperature (25 °C) and after being thermally treated at 300 °C, compared with the thermal treatment at the other temperatures. Figure 9 mainly shows the morphologies of the GYP, TAC, and SIL specimens without thermal treatment (at 25 °C) and after being treated at 300 °C. At 25 °C, the microstructure of the GYP specimen demonstrated needle-shaped crystals bonded together into various clusters distributed among the whole matrix. The bonding between single crystals or crystal clusters seemed negatively influenced (discontinued) by the matrix's gel and capillary pores [39,40]. The TAC specimen demonstrated a gypsum matrix different from the GYP specimen. The morphology of the gypsum crystal was quite varied in diameter and length, with apparently weak bonding among single crystals. The crystal clusters seemed not to have formed in this matrix, and, meanwhile, the interface between the talc and gypsum crystals was discontinuous with some gaps. The matrix appeared loosely ordered and organized, lacked consistent crystals and tight integration, and contained apparent voids and relatively big pores. The SIL specimen presented a well-developed matrix of gypsum crystals integrated firmly with silica flour particles. The morphology of each gypsum crystal appeared similar to that of the GYP specimen. Still, the cluster formed by bonded crystals looked more consistent and well-distributed than that in the GYP specimen. The interface between silica flour particles and gypsum crystals was continuous without showing apparent gaps or voids. The whole matrix was compact, consistent, and bonded without showing problematic porosity, partially due to silica flour particles' filling capacity. The microstructural settings of the three gypsum matrixes agreed well with their mechanical properties observed at the macroscopic level.

Figure 9. Electron microscope images of GYP, TAC, and SIL specimens before (25 °C, **left** column) and after the thermal treatment at 300 °C (**right** column).

After being treated at 300 °C, as shown in Figure 9 (right column), the gypsum phases in the matrix turned into gypsum hemihydrate and anhydrite, which was also confirmed by the XRD results in Figure 8. For the TAC and SIL specimens, the talc and quartz remained intact in the matrix after being treated at 300 °C. For the GYP specimen, the thermal treatment generally reduced its mechanical strength due to phase change and pore formation, as shown in the SEM images. Although the mechanical strength was better than at 100 °C and 200 °C after being treated at 300 °C owing to the formation of anhydrite, its strength was still lower than at 25 °C (without treatment). However, in the TAC and SIL specimens, the strength after being treated at 300 °C was notably increased compared with the treatments at 100 °C and 200 °C and without treatment (25 °C). This was related to the combined gypsum phase change and the filling or skeletal effects of talc and silica flour particles in the gypsum matrix at the microscopic level.

4. Discussion

Adding talc powder or silica flour to gypsum reduced the hydration heat flow of the gypsum blends. Talc powder was more effective in delaying gypsum hydration, allowing extended sample casting and preparation time. Replacing 10% of gypsum with talc powder decreased the compressive strength of blended gypsum at 25 °C. Still, it reduced the thermal conductivity of the blend by 30% compared to the GYP (pure gypsum) specimen. Replacing 10% of gypsum with silica flour increased the compressive strength of blended gypsum at 25 °C. Also, the thermal conductivity of the blended gypsum was reduced by more than 60% compared to the GYP specimen. Silica flour seemed more effective than talc powder in modifying blended gypsum's strength and thermal conductivity. The TAC specimen had higher weight loss and shrinkage than the SIL specimen during the thermal treatment, likely due to talc powder's capacity to hold water on its particle surfaces more robustly than silica flour particles.

Silica flour was superior to talc powder in blended gypsum because the SIL specimen developed higher compressive strength, lower thermal conductivity, less weight loss, and smaller shrinkage than the TAC specimen. The weight loss of all gypsum specimens continued to increase with the increase in thermal treatment temperature. This was mainly due to water being driven from the gypsum matrix at temperatures up to about 300 °C and gases (such as CO_2 and SO_2) from the decomposition of relevant minerals at higher treatment temperatures being released. The samples' thermal conductivity decreased with the treatment temperature up to 200 °C. The results at 300 °C and beyond were unavailable due to significant cracking in the specimens. It was interesting to see that, at 100 °C, the TAC specimen exhibited a negative shrinkage (slight expansion), resulting from the evaporation of water in the vicinity of talc powder particles that possibly rehydrated the relevant gypsum phases during the same period.

The compressive strength of hardened gypsum decreased after being treated at 100 °C and 200 °C, but it increased again after being treated at 300 °C. Such a strength increase was even more significant in the blended gypsum specimens with talc powder or silica flour. Anhydrite and gypsum hemihydrate were the two primary phase assemblages in the gypsum matrix after being treated at 300 °C. Anhydrite was likely formed first during the treatment, while a portion was rehydrated with water vapor into gypsum hemihydrate during the cooling process. Anhydrite was a more robust phase from a strength point of view than the other gypsum phases. The filler and skeletal functions provided by talc powder or silica flour in the matrix facilitated the strength increase after the 300 °C treatment. Meanwhile, the microcracking developed in the matrix during the heat treatment might have contributed to energy dissipation and stress redistribution. This required further investigation regarding the temperature-dependent mechanism since the cracking intensified at higher temperatures, so the specimens became weaker against the external load.

5. Conclusions

This work investigated mineral additions (talc powder and silica flour) to gypsum and their influences on blended gypsum's physical and mechanical properties at elevated temperatures. The experimental results and discussions indicate the following conclusions:

(1) The addition of silica flour (10% replacement of gypsum) improved gypsum's physical and mechanical properties, such as compressive strength, thermal conductivity, and shrinkage at elevated temperatures. Silica flour performed better than talc powder (at the same dosage) in achieving blended gypsum's physical and mechanical properties in this work. It is suggested that silica flour can be added to gypsum for better performance in normal temperatures and high-temperature resistance. The addition of silica powder could fill in some micropores, thereby increasing the density of the gypsum board.

(2) The thermal treatment at 100 °C and 200 °C decreased the mechanical properties of gypsum paste owing to the loss of water, phase change, and formation of pores.

However, the mechanical strength of gypsum paste recovered to some extent after being treated at 300 °C due to the formation of a more robust anhydrite phase. With the addition of talc powder or silica flour, the strength regained in the gypsum paste after being treated at 300 °C was remarkable, owing to the filler and skeletal effects of the talc powder and silica flour particles in the gypsum matrix combined with gypsum's phase change.

(3) The thermal-induced cracking seemed to cause no structural damage to the specimens when this work's thermal treatment temperature was lower than 300 °C. To a certain extent, microcracks can enhance the ability of gypsum boards to absorb more energy. However, further investigations were needed to explore the relationship between thermal-induced cracking and mechanical properties. Both silica flour and talc powder demonstrated no apparent roles in controlling thermal-induced crack formation in the gypsum matrix.

Author Contributions: Conceptualization, J.W. and E.L.; Methodology, J.W. and E.L.; Software, J.W.; Validation, J.W.; Formal analysis, J.W.; Investigation, J.W.; Resources, E.L.; Data curation, J.W.; Writing—original draft, J.W. and E.L.; Writing—review & editing, J.W. and E.L.; Supervision, E.L.; Project administration, E.L.; Funding acquisition, E.L. All authors have read and agreed to the published version of the manuscript.

Funding: This work has been funded by the Xi'an Jiaotong-Liverpool University Research Development Funding (RDF-19-01-22) and by the Departments of Civil Engineering at Xi'an Jiaotong-Liverpool University and Tsinghua University.

Institutional Review Board Statement: Not applicable.

Informed Consent Statement: Not applicable.

Data Availability Statement: Data are contained within the article.

Conflicts of Interest: The authors declare no conflict of interest.

References

1. Monteiro, P.J.; Miller, S.A.; Horvath, A. Towards sustainable concrete. *Nat. Mater.* **2017**, *16*, 698–699. [CrossRef] [PubMed]
2. Xu, L.; Wang, J.J.; Li, K.F.; Lin, S.Y.; Li, M.L.; Hao, T.Y.; Ling, Z.; Xiang, D.; Wang, T.Y. A systematic review of factors affecting properties of thermal-activated recycled cement. *Resour. Conserv. Recycl.* **2022**, *185*, 106432. [CrossRef]
3. Murat, M.; Attari, A. Modification of some physical properties of gypsum plaster by addition of clay minerals. *Cem. Concr. Res.* **1991**, *21*, 378–387. [CrossRef]
4. Lushnikova, N.; Dvorkin, L. *Sustainability of Gypsum Products as a Construction Material, Sustainability of Construction Materials*; Elsevier: Amsterdam, The Netherlands, 2016; pp. 643–681.
5. Khaliullin, M.; Dimieva, A. Composite gypsum binder under introducing thermally activated clay as a pozzolanic component and adding ground limestone. *IOP Conf. Ser. Mater. Sci. Eng.* **2020**, *890*, 012093. [CrossRef]
6. Sophia, M.; Sakthieswaran, N. Synergistic effect of mineral admixture and bio-carbonate fillers on the physico-mechanical properties of gypsum plaster. *Constr. Build. Mater.* **2019**, *204*, 419–439. [CrossRef]
7. Khaliullin, M.; Gilmanshina, A. The effect of additives of mechanically activated mineral fillers on the properties of composite gypsum binders. In Proceedings of the International Scientific Conference on Socio-Technical Construction and Civil Engineering, Kazan, Russia, 21–28 April 2021; Springer: Cham, Switzerland, 2021; pp. 334–342.
8. Lanzón, M.; Castellón, F.J.; Ayala, M. Effect of the expanded perlite dose on the fire performance of gypsum plasters. *Constr. Build. Mater.* **2022**, *346*, 128494. [CrossRef]
9. Huang, Y.; Xu, C.; Li, H.; Jiang, Z.; Gong, Z.; Yang, X.; Chen, Q. Utilization of the black tea powder as multifunctional admixture for the hemihydrate gypsum. *J. Clean. Prod.* **2019**, *210*, 231–237. [CrossRef]
10. Lesovik, V.; Chernysheva, N.; Fediuk, R.; Amran, M.; Murali, G.; de Azevedo, A.R. Optimization of fresh properties and durability of the green gypsum-cement paste. *Constr. Build. Mater.* **2021**, *287*, 123035. [CrossRef]
11. Doleželová, M.; Scheinherrová, L.; Krejsová, J.; Vimmrová, A. Effect of high temperatures on gypsum-based composites. *Constr. Build. Mater.* **2018**, *168*, 82–90. [CrossRef]
12. Bicer, A.; Kar, F. Thermal and mechanical properties of gypsum plaster mixed with expanded polystyrene and tragacanth. *Therm. Sci. Eng. Prog.* **2017**, *1*, 59–65. [CrossRef]
13. Gencel, O.; del Coz Diaz, J.J.; Sutcu, M.; Koksal, F.; Rabanal, F.P.Á.; Martinez-Barrera, G. A novel lightweight gypsum composite with diatomite and polypropylene fibers. *Constr. Build. Mater.* **2016**, *113*, 732–740. [CrossRef]

14. Camarini, G.; Pinto, M.C.C.; de Moura, A.G.; Manzo, N.R. Effect of citric acid on properties of recycled gypsum plaster to building components. *Constr. Build. Mater.* **2016**, *124*, 383–390. [CrossRef]
15. Lanzón, M.; García-Ruiz, P. Effect of citric acid on setting inhibition and mechanical properties of gypsum building plasters. *Constr. Build. Mater.* **2012**, *28*, 506–511. [CrossRef]
16. Khalil, A.; Tawfik, A.; Hegazy, A. Plaster composites modified morphology with enhanced compressive strength and water resistance characteristics. *Constr. Build. Mater.* **2018**, *167*, 55–64. [CrossRef]
17. Pundir, A.; Garg, M.; Singh, R. Evaluation of properties of gypsum plaster-superplasticizer blends of improved performance. *J. Build. Eng.* **2015**, *4*, 223–230. [CrossRef]
18. Tokarev, Y.; Ginchitsky, E.; Sychugov, S.; Krutikov, V.; Yakovlev, G.; Buryanov, A.; Senkov, S. Modification of gypsum binders by using carbon nanotubes and mineral additives. *Procedia Eng.* **2017**, *172*, 1161–1168. [CrossRef]
19. Zhu, C.; Zhang, J.; Peng, J.; Cao, W.; Liu, J. Physical and mechanical properties of gypsum-based composites reinforced with PVA and PP fibers. *Constr. Build. Mater.* **2018**, *163*, 695–705. [CrossRef]
20. Azad, A.; Mousavi, S.-F.; Karami, H.; Farzin, S. Application of talc as an eco-friendly additive to improve the structural behavior of porous concrete, Iranian Journal of Science and Technology. *Trans. Civ. Eng.* **2019**, *43*, 443–453.
21. Berodier, E.; Scrivener, K. Understanding the Filler Effect on the Nucleation and Growth of C-S-H. *J. Am. Ceram. Soc.* **2014**, *97*, 3764–3773. [CrossRef]
22. Tavares, L.R.C.; Junior, J.F.T.; Costa, L.M.; da Silva Bezerra, A.C.; Cetlin, P.R.; Aguilar, M.T.P. Influence of quartz powder and silica fume on the performance of Portland cement. *Sci. Rep.* **2020**, *10*, 21461. [CrossRef]
23. Rashad, A.M. Effect of quartz-powder on the properties of conventional cementitious materials and geopolymers. *Mater. Sci. Technol.* **2018**, *34*, 2043–2056. [CrossRef]
24. Ahmed, A.; Mahmoud, A.A.; Elkatatny, S. The Use of the Granite Waste Material as an Alternative for Silica Flour in Oil-Well Cementing. *ACS Omega* **2020**, *5*, 32341–32348. [CrossRef] [PubMed]
25. Liu, X.; Liu, X.; Hu, Y. Investigation of the thermal decomposition of talc. *Clays Clay Miner.* **2014**, *62*, 137–144. [CrossRef]
26. Wentzensen, N.; O'Brien, K.M. Talc, body powder, and ovarian cancer: A summary of the epidemiologic evidence. *Gynecol. Oncol.* **2021**, *163*, 199–208. [CrossRef]
27. Carretero, M.I.; Pozo, M. Clay and non-clay minerals in the pharmaceutical industry: Part I. Excipients and medical applications. *Appl. Clay Sci.* **2009**, *46*, 73–80. [CrossRef]
28. Xu, L.; Wang, J.J.; Li, K.F.; Hao, T.Y.; Li, Z.; Li, L.; Ran, B.; Du, H. New insights on dehydration at elevated temperature and rehydration of GGBS blended cement. *Cem. Concr. Compos.* **2023**, *139*, 105068. [CrossRef]
29. Xu, L.; Wang, J.J.; Li, K.F.; Li, M.L.; Lin, S.Y.; Hao, T.Y.; Wang, T.Y.; Guo, Y.P.; Ling, Z. Investigations on the rehydration of recycled blended SCMs cement. *Cem. Concr. Res.* **2023**, *163*, 107036. [CrossRef]
30. Zhang, L.; Liu, C.; Liu, L.; Zhang, H. Study on early hydration of gypsum-based materials containing different chemical admixtures by isothermal calorimetry and oscillation rheology. *J. Therm. Anal. Calorim.* **2022**, *147*, 6099–6107. [CrossRef]
31. Magallanes-Rivera, R.; Escalante-Garcia, J.; Gorokhovsky, A. Hydration reactions and microstructural characteristics of hemihydrate with citric and malic acid. *Constr. Build. Mater.* **2009**, *23*, 1298–1305. [CrossRef]
32. Yang, L.; Shi, C.; Liu, J.; Wu, Z. Factors affecting the effectiveness of internal curing: A review. *Constr. Build. Mater.* **2021**, *267*, 121017. [CrossRef]
33. Baux, C.; Melinge, Y.; Lanos, C.; Jauberthie, R. Enhanced gypsum panels for fire protection. *J. Mater. Civ. Eng.* **2008**, *20*, 71–77. [CrossRef]
34. Rahmanian, I.; Wang, Y. A combined experimental and numerical method for extracting temperature-dependent thermal conductivity of gypsum boards. *Constr. Build. Mater.* **2012**, *26*, 707–722. [CrossRef]
35. Ritterbach, L.; Becker, P. Temperature and humidity dependent formation of CaSO4· xH2O (x= 0... 2) phases. *Glob. Planet. Change* **2020**, *187*, 103132. [CrossRef]
36. Borrachero, M.; Paya, J.; Bonilla, M.; Monzó, J. The use of thermogravimetric analysis technique for the characterization of construction materials: The gypsum case. *J. Therm. Anal. Calorim.* **2008**, *91*, 503–509. [CrossRef]
37. Piga, L.; Villieras, F.; Yvon, J. Thermogravimetric analysis of a talc mixture. *Thermochim. Acta* **1992**, *211*, 155–162. [CrossRef]
38. Tang, Y.; Gao, J.; Liu, C.; Chen, X.; Zhao, Y. Dehydration pathways of gypsum and the rehydration mechanism of soluble anhydrite γ-$CaSO_4$. *ACS Omega* **2019**, *4*, 7636–7642. [CrossRef]
39. He, H.; Wang, Y.; Wang, J. Effects of aggregate micro fines (AMF), aluminum sulfate and polypropylene fiber (PPF) on properties of machine-made sand concrete. *Appl. Sci.* **2019**, *9*, 2250. [CrossRef]
40. Wang, Y.; Lu, H.; Wang, J.; He, H. Effects of highly crystalized nano C-S-H particles on performances of Portland cement paste and its mechanism. *Crystals* **2020**, *10*, 816. [CrossRef]

coatings

Article

Novel Perimidine Derivatives as Corrosion Inhibitors of HRB400 Steel in Simulated Concrete Pore Solution

Qi Ma [1,2], Jingshun Cai [1,2], Song Mu [1,2,*], Hao Zhang [1,2], Kai Liu [1,2], Jianzhong Liu [1,2] and Jinxiang Hong [1,2]

1 State Key Laboratory of High Performance Civil Engineering Materials, Jiangsu Research Institute of Building Science Co., Ltd., Nanjing 211103, China
2 Jiangsu Sobute New Materials Co., Ltd., Nanjing 211103, China
* Correspondence: musong@cnjsjk.cn

Abstract: 1H-Perimidine (PMD) and 1H-perimidine-2-thiol (SPMD) were developed as inhibitors for reinforcing steel in a simulated concrete pore (SCP) solution. Electrochemical measurements, contact angle experiments, scanning electron microscopy (SEM) and quantum chemical calculations were used to investigate the corrosion performance and adsorption mechanisms. The experimental results showed that owing to the structure of SPMD containing a 2-position sulfhydryl group, SPMD is superior to PMD as a corrosion inhibitor for HRB400 reinforced steel in the SCP solution, and its corrosion efficiency can reach more than 80%. Moreover, the introduction of nitrogen and sulfur atoms into the inhibitor not only can coordinate with Fe atoms to form strong bonds but also is useful for preventing charge transfer in the metal corrosion process. More importantly, the perimidine derivatives can spontaneously adsorb on iron, and the adsorption process obeys the Langmuir isotherm. The research results show that the perimidine derivatives can improve the durability of concrete structures.

Keywords: perimidine derivatives; inhibitor; simulated concrete pore solution; HRB400 steel; adsorption mechanism

1. Introduction

In a chloride salt and marine environment, the reinforcement in concrete is more vulnerable to corrosion by chlorine ions, which quickens the corrosion of steel and then shortens the service life of the concrete structure [1,2]. This not only has hidden social security dangers but likewise causes huge economic losses for the country [3,4]. Accordingly, it is of great significance to solve the corrosion problem of steels in concrete. There are numerous approaches to averting the corrosion of steels in concrete, such as concrete surface coating, epoxy reinforcement, stainless steel reinforcement, cathodic protection, electrochemical chloride removal and corrosion inhibitors for reinforcement, etc. [5]. Due to the simple application, low cost and remarkable rust resistance effect of corrosion inhibitors, which are widely used in the concrete industry, people are paying more and more attention to the research of inhibitors for reinforced steel in concrete [6]. Currently, there are three types of corrosion inhibitors: inorganic, organic and mixed corrosion inhibitors [7]. The inorganic inhibitors are mainly nitrite. On the one hand, nitrite has a poor resistance effect. On the other hand, it has carcinogenic effects. Hence, many countries have banned the use of nitrite in the concrete field [8,9].

In recent years, numerous people have begun to study the synthesis of organic corrosion inhibitors with simple methods and good rust resistance effects. The efficiency of a corrosion inhibitor is linked to its structure [10]. The corrosion inhibition mechanism mainly includes physical adsorption, chemical adsorption and π bond adsorption. Generally, heteroatoms such as sulfur, phosphorus, oxygen, nitrogen and π electrons interact with the empty d orbital of iron to form a coordination bond [11]. Therefore, the inhibitors are adsorbed on the iron surface, showing good inhibition efficiency [12]. For example, naphthalene and pyrimidine derivatives, which are electron-rich aromatic and heteroaromatic compounds, have been proven to be superior corrosion inhibitors in corrosive media [13]. It is predicted that perimidine derivatives can play an effective inhibitory role in aggressive media because their naphthalene rings are fused with pyrimidine rings. In this study, 1*H*-perimidine (PMD) and 1*H*-perimidine-2-thiol (SPMD) were prepared, and their corrosion properties on metal in a simulated concrete pore (SCP) solution were investigated. The mechanism of perimidine derivatives adsorbing on the iron surface was equally further studied through theoretical chemical calculations.

2. Materials and Methods

2.1. Synthesis of Inhibitors

PMD and SPMD were synthesized through a modified procedure reported previously [14,15], as shown in Figure 1.

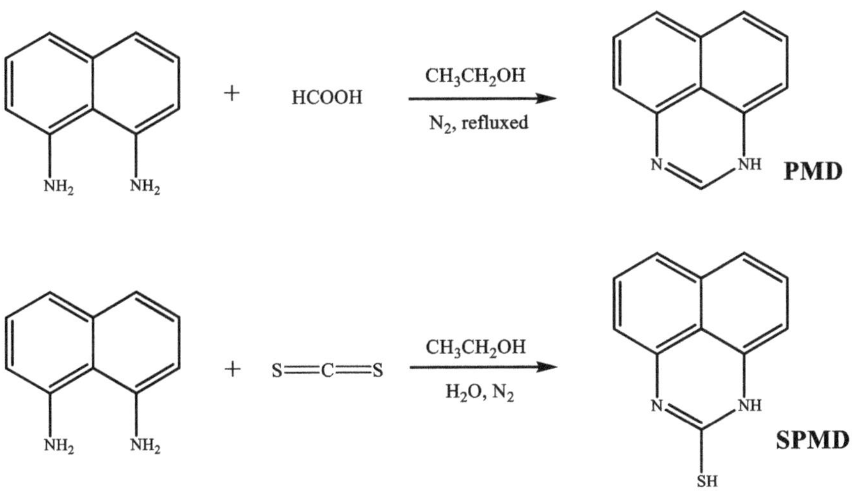

Figure 1. Synthesis routes of PMD and SPMD.

2.1.1. 1H-Perimidine (PMD)

To naphthalene-1,8-diamine (7.49 g, 47.40 mmol) in ethanol (25 mL), formic acid (10.53 mL, 0.27 mol) was added. The resulting mixture was refluxed under N_2 for 2 h, and then the reaction mixture was diluted with water and alkalized with NH_4OH (17.3%). The sediment was dried under vacuum. The product was recrystallized from ethanol to obtain yellow crystals. Yield 6.53 g (87 %). ^{1}H-NMR (400 MHz, D_2O) δ ppm: 8.43 (s, 1H), 7.37 (br s, NH), 7.05 (d, 2H), 6.92 (t, 2H), 6.05 (d, 2H). IR (KBr) *v*: 3430, 2914, 1628, 1503, 1379, 1039.

2.1.2. 1H-Perimidine-2-thiol (SPMD)

To naphthalene-1,8-diamine (5.93 g, 37.50 mmol) in 95% aqueous ethanol (30 mL) solution, KOH (15.00 mg, 0.27 mmol) and CS_2 (2.70 mL, 45.00 mmol) were added under a nitrogen atmosphere. Precipitation of the product began almost immediately, and after 30 min, the mixture was filtered to give 1*H*-perimidine-2-thiol as a gray solid. Yield 6.76 g (90%). ^{1}H-NMR (400 MHz, DMSO-d6) δ ppm: 10.35 (s, 1H, NH), 7.82 (t, 2H), 7.59 (t, 2H), 7.43 (d, 2H), 2.88 (s, 1H, SH). IR (KBr) *v*: 3425, 2944, 2543, 1605, 1510, 1385, 1042.

2.2. Materials

The chemicals were analytical grade (>98.5% pure). The specimens were HRB400 reinforced steel containing 0.24 *wt.*% C, 1.6 *wt.*% Mn, 0.75 *wt.*% Si, 0.05 *wt.*% S, 0.045% P and balance iron. The test solution used in all experiments was a simulated concrete pore (SCP) solution, which was 3.5% NaCl saturated with a $Ca(OH)_2$ solution (pH 12.5). Ultrapure water was used to prepare the experimental solution for electrochemical measurements. Different concentrations (0.02 to 0.12 mM) of corrosion inhibitors were prepared in the simulated concrete pore solution.

2.3. Electrochemical Measurements

Electrochemical impedance spectroscopy (EIS) measurements were performed using a PARSTAT 4000 electrochemical workstation. A typical three-electrode cell was used, equipped with a platinum electrode as the auxiliary electrode and a Ag/AgCl, KClsat electrode as the reference electrode. An epoxy-encapsulated HRB400 reinforced electrode with an exposed area of 1.0 cm^2 was employed as the working electrode. Before the test, the HRB400 reinforced electrode was ground with different sandpapers (grades 500, 800, 1000, 1500 and 2000), cleaned with distilled water and ethanol for 3 min, and finally dried under N_2 for standby use. The HRB400 reinforced electrode was submersed in the SCP solution with perimidine derivatives. After three days, potentiodynamic polarization curves were measured in a range of −0.5 V to 1.0 V with a scan rate of 0.5 mV/s. EIS tests were conducted at 25 °C in the three-electrode cell. The measurement was conducted at open-circuit potential (OCP) over the frequency range of 100 kHz–10^{-2} Hz with an amplitude of 10 mV. The test data were analyzed by the software of the PARSTAT 4000 electrochemical workstation.

2.4. Surface Morphological Test and Contact Angle Analysis

HRB400 reinforced steel was submersed in the SCP solution with perimidine derivatives. The steel specimens were immediately taken out for three days, the corrosion products on the steels were removed with distilled water and anhydrous ethanol successively, and then scanning electron microscope measurements were conducted using a Quanta 250 instrument(Thermo Fisher Scientific, Waltham, MA, USA). Contact angles of specimens exposed to the SCP solution without and with 0.12 mM PMD and SPMD were measured using a DSA255 Contact Angle Tester (KRUSS, Hamburg, Germany). with a 2 mL liquid drop.

2.5. Quantum Chemical Calculations

GAMESS-US software was used to the study of the perimidine derivative molecules using density functional theory (DFT), with the B3LYP functional under the 6-31G (d) basis sets [16,17]. The removal of imaginary frequencies was carried out to confirm that the optimized geometries are true minima. The Debye (μ), lowest unoccupied molecular orbital energy (E_{LUMO}) and highest occupied molecular orbital energy (E_{HOMO}) were calculated using the software. Other electronic properties were obtained by the following equations:

$$\Delta E = E_{LUMO} - E_{HOMO} \tag{1}$$

$$A = -E_{LUMO} \tag{2}$$

$$I = -E_{HOMO} \tag{3}$$

$$\chi = (I + A)/2 \tag{4}$$

$$\eta = (I - A)/2 \tag{5}$$

where ΔE is the gap energy between the HOMO and LUMO, A is electronic affinity, I is ionization potential, χ is electronegativity and η is chemical hardness. In addition, the transferred electrons (ΔN) can be computed from Equation (6) [18]:

$$\Delta N = \frac{\chi_{Fe} - \chi_{inh}}{2(\eta_{Fe} + \eta_{inh})} \tag{6}$$

where χ_{Fe} is 7 eV, χ_{inh} is the electronegativity of the inhibitor, η_{inh} is the absolute electronegativity of the inhibitor, and η_{Fe} is the absolute electronegativity of iron, which is 0 eV [19].

3. Results and Discussion

3.1. Electrochemical Experiments

3.1.1. Potentiodynamic Polarization (PDP)

Figure 2 shows the potentiodynamic polarization curves of reinforced HRB400 steel in the SCP solution at 25 °C. From the polarization curve, the corrosion potential (E_{corr}), the pitting potential (E_{pit}), the corrosion current density (I_{corr}), cathodic and anodic Tafel slopes (β_c and β_a) and inhibition efficiency (IE) can be obtained [20]. All parameters are listed in Table 1. The inhibition efficiency is expressed as [21]:

$$IE = \frac{I^0_{corr} - I_{corr}}{I^0_{corr}} \times 100\% \quad (7)$$

where I^0_{corr} and I_{corr} are the current densities of the working electrode without and with a corrosion inhibitor.

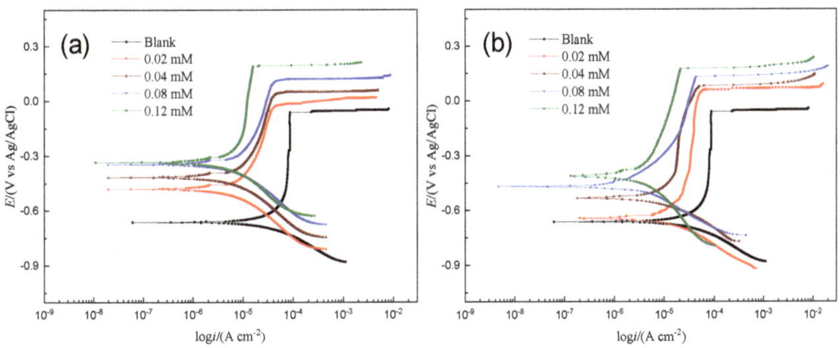

Figure 2. Potentiodynamic polarization for the corrosion of HRB400 reinforced steel in SCP solution with different concentrations of PMD (**a**) and SPMD (**b**) at 25 °C.

Table 1. PDP parameters for HRB400 reinforced steel in SCP solution.

Inhibitors	Conc.	E_{corr}	E_{pit}	$-\beta_c$	β_a	I_{corr}	IE	θ
	mM	mv	mv	mV dec^{-1}	mV dec^{-1}	μA cm^{-2}	%	
	Blank	−658.81 ± 5.50	−73.66 ± 3.15	197.24 ± 3.96	251.39 ± 4.56	23.69 ± 1.15	/	/
PMD	0.02	−486.62 ± 4.75	−26.89 ± 2.78	177.24 ± 4.51	196.00 ± 4.12	12.97 ± 1.01	45.25	0.4525
	0.04	−415.19 ± 6.23	55.69 ± 1.96	168.08 ± 3.62	194.10 ± 4.37	9.28 ± 1.37	60.84	0.6084
	0.08	−350.40 ± 5.66	132.46 ± 3.05	170.54 ± 3.74	187.13 ± 3.95	6.24 ± 0.86	73.66	0.7366
	0.12	−330.47 ± 4.97	194.15 ± 2.94	164.85 ± 3.55	189.15 ± 4.16	3.68 ± 0.65	84.48	0.8448
SPMD	0.02	−435.85 ± 7.32	53.94 ± 1.75	167.17 ± 4.09	187.31 ± 5.03	10.43 ± 1.27	55.99	0.5599
	0.04	−408.14 ± 6.40	91.32 ± 2.46	170.88 ± 4.27	190.54 ± 4.48	7.36 ± 0.79	68.95	0.6895
	0.08	−343.98 ± 5.73	130.06 ± 3.54	166.43 ± 3.05	193.39 ± 3.72	5.07 ± 0.93	78.60	0.7860
	0.12	−315.49 ± 5.49	175.99 ± 3.17	163.39 ± 3.96	186.27 ± 3.85	2.94 ± 0.82	87.61	0.8761

In general, as the concentration of PMD or SPMD increases, the corrosion current density gradually decreases in the SCP solution. In other words, as the concentrations of inhibitors increase, the inhibitors gradually adsorb on iron. Therefore, iron can effectively prevent corrosion from chloride ions. For the anodic curve, the addition of inhibitors significantly inhibited the dissolution of HRB400 reinforcement in the SCP, and with the increase in the perimidine derivative concentration, the inhibition of iron anode dissolution was greater, and the shape of the Tafel curve remained unchanged. The results show that the process of the anode inhibiting the iron reaction has not changed [22,23]. The cathodic curve also effectively inhibited corrosion and prevented the oxygen reduction reaction. In addition, the inhibitory effect was more pronounced in the anode part, indicating that the perimidine derivatives are inhibitor molecules with a better anodic inhibition performance.

The electrochemical parameters for HRB400 reinforced steel in the SCP solution with different concentrations of PMD (a) and SPMD (b) at 25 °C are presented in Table 1. The surface coverage (θ) is equal to the inhibition efficiency (*IE*). It can be seen from Table 1 that with the increase in the perimidine derivative concentration, the corrosion potential gradually shifts positively, and the potential shifts by more than 85 mV after adding the inhibitor; the results indicate that the two perimidine derivatives PMD and SPMD are anodic corrosion inhibitors [24–26]. The pitting potential is the potential at which the anode current rapidly increases to a higher value when the surface passivation film breaks. The results show that the pitting potential increases gradually with the increase in the perimidine derivative concentration, suggesting that the inhibitors can inhibit the corrosion of chloride ions on the surface of the reinforced steel. This can be attributed to perimidine derivative molecules and chloride ions competing for the surface of the specimen, and the inhibitor is adsorbed on the metal to prevent the adsorption of chloride ions onto steel surfaces [27,28]. Moreover, after adding the inhibitor, the positive corrosion potential shift will reduce the interaction between chlorine ions and active sites on the iron surface, indicating that greater activation energy is needed if pitting corrosion occurs between chlorine ions and the steel, thus improving the pitting resistance of the sample [29,30].

Furthermore, as the concentration of the perimidine derivative increases, the Icorr value gradually decreases and the surface coverage gradually increases, indicating that the corrosion inhibitory effect is getting better. This effect can be ascribed to the perimidine derivative gradually occupying the active centers of the iron surface and gradually covering the iron surface. Furthermore, the Icorr value of SPMD was smaller than that of PMD, suggesting that SPMD had a better inhibitory effect than PMD.

3.1.2. EIS

Figure 3 presents the Nyquist plots and Bode plots for HRB400 reinforced steel in the SCP solution with the perimidine derivatives at 25 °C. In the blank SCP solution, pitting corrosion occurred on iron due to corrosion by Cl^-. As the immersion time increases, the surface of the steel is more seriously corroded. Therefore, the Nyquist plots show lower capacitive resistance. After adding the perimidine derivative, it can be adsorbed on iron to form an adsorption film to resist erosion by chlorine ions and protect the steel. As the concentration of the perimidine derivative increases, the protective film becomes more complete and dense. Hence, chlorine ions have more difficulty coming into contact with the steel. These results suggest that with the increase in the concentration of the perimidine derivative, the Nyquist capacitive reactance becomes larger and the inhibitory effect becomes stronger, which can also be obtained by analyzing the Bode diagram. It can be observed from the EIS in Figure 3 that the capacitive reactance of SPMD is larger than that of PMD at the same concentration, which suggests that the *IE* of SPMD is better than that of PMD in the SCP solution.

In addition, the capacitive reactance of Nyquist plots is slightly depressed as a semicircular shape in the absence and presence of perimidine derivatives, where the deformation of the capacitance semicircle is attributed to surface roughness or unevenness [31]. Figure 3 shows that the electrochemical impedance spectroscopy has two characteristics of time constants in the SCP solution without and with inhibitors. These results reveal that the first time constant is related to the properties of the charged double electric layer; that is, charge transfer or material transfer occurs in the corrosion process [32]. A high frequency is related to the protective film on the iron, which is the film formed by inhibitor molecules, iron oxide and hydroxide [33]. The electrochemical parameters were obtained by the equivalent circuit in Figure 4. In the equivalent circuit diagram, R_S is the solution resistance, R_f is the passivation/adsorption film resistance and R_{ct} is thc charge transfer resistance. CPE_1 and CPE_2 are the double electric layer and the adsorption film constant phase element, respectively. In order to obtain a fitting curve that is more consistent with the experimental data, the constant phase component CPE is used in the equivalent circuit to replace the pure capacitor, thereby obtaining more reliable fitting data. The constant phase element (Z_{CPE}) is shown in the following equation [34]:

$$Z_{CPE} = Y_0^{-1}(j\omega)^{-n} = Y_0^{-1}\omega^{-n}\left(\cos\frac{n\pi}{2} - j\sin\frac{n\pi}{2}\right) \quad (8)$$

where Y_0 is a proportional coefficient, ω is the angular frequency, n is the phase shift, j is the imaginary number and $j^2 = -1$. For $n = -1$, CPE is a pure inductance element. For $n = 0$, CPE is a pure resistance element. For $n = 0.5$, CPE is a Warburg element, and for $n = 1$, CPE is a pure capacitor element. The value of n depends on the non-uniformity and roughness of the metal surface. The inhibition efficiency (IE) is given as:

$$IE(\%) = \frac{R_{ct} - R_{ct}^0}{R_{ct}} \times 100 \quad (9)$$

where R_{ct}^0 and R_{ct} are the charge transfer resistance of the steel electrode without and with the perimidine derivatives.

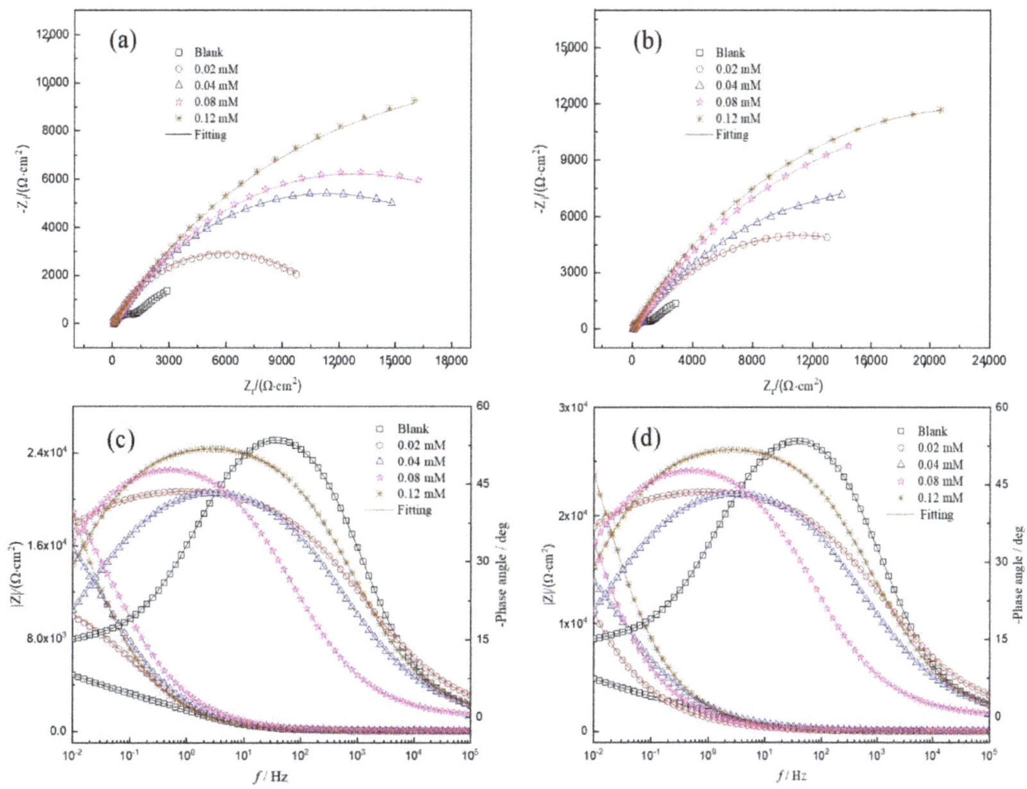

Figure 3. EIS plots for reinforced steel in SCP with perimidine derivatives at 25 °C. Top: Nyquist diagrams; bottom: Bode diagrams. PMD (**a**,**c**); SPMD (**b**,**d**).

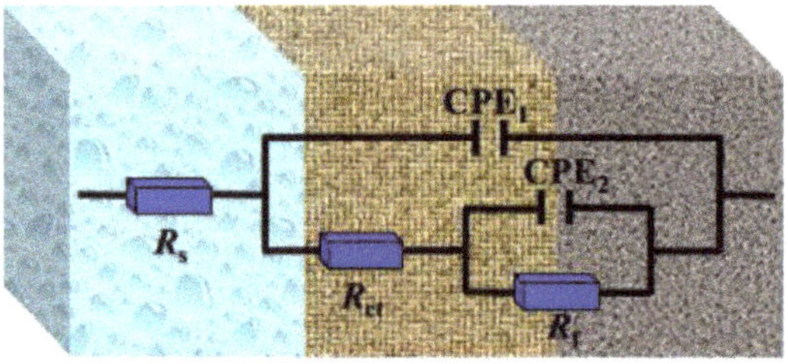

Figure 4. Equivalent circuit used to fit the impedance spectra.

The EIS parameters are displayed in Table 2, which could be obtained by the equivalent circuit. It was observed that with the increase in the perimidine derivative concentration, the Y_0 value in CPE_1 and CPE_2 gradually decreased, where the decrease in Y_0 in CPE_1 indicates that inhibitor molecules gradually replace water and other substances on the surface of reinforcement, leading to a decrease in the dielectric constant or an increase in the thickness of the double electric layer, which reduces the double-layer capacitance in turn. In addition, corrosion inhibitor molecules gradually adsorb onto iron, resulting in a decrease in film capacitance (CPE_2) and an increase in film resistance (R_f) and charge transfer resistance (R_{ct}) [35], thereby inhibiting the erosion effect of chloride ions on the reinforcement. On the one hand, inhibitors are adsorbed on iron, reducing chloride ion adsorption on active sites [36] and reducing pitting corrosion and local rust. On the other hand, inhibitor molecules adsorb to form a film on the reinforcement, and with the increase in the perimidine derivative concentration, the protective film becomes more complete and dense so as to prevent erosion by chloride ions. Furthermore, both PMD and SPMD at a concentration of 0.12 mM can obviously inhibit corrosion by chlorine ions in the SCP solution. It is worth noticing that SPMD has better *IE* than PED at the same concentration. In other words, SPMD has better corrosion resistance than PED, which is also consistent with the results of the potentiodynamic polarization test.

Table 2. EIS parameters of reinforced steel in SCP solution with perimidine derivatives.

C_{inh}/(mM)		R_s	CPE1		R_f	CPE_2		R_{ct}	*IE*
		Ω cm^2	Y_0 (μΩ^{-1}S^{-n}cm^{-2})	n_1	kΩ cm^2	Y_0 (μΩ^{-1}S^{-n}cm^{-2})	n_2	kΩ cm^2	%
	Blank	9.89 ± 1.01	410.9 ± 3.5	0.89 ± 0.03	0.41 ± 0.04	1 378.1 ± 10.5	0.61 ± 0.05	6.94 ± 0.52	/
PED	0.02	30.05 ± 2.25	306.5 ± 3.0	0.93 ± 0.05	0.59 ± 0.03	783.5 ± 10.8	0.65 ± 0.03	11.95 ± 0.69	41.92
	0.04	64.03 ± 2.53	247.1 ± 2.2	0.84 ± 0.07	1.32 ± 0.05	512.2 ± 6.2	0.59 ± 0.04	17.01 ± 0.83	59.20
	0.08	75.18 ± 2.45	179.8 ± 2.3	0.88 ± 0.08	2.01 ± 0.02	405.6 ± 8.5	0.56 ± 0.05	25.28 ± 0.75	72.55
	0.12	82.95 ± 2.54	105.2 ± 2.0	0.90 ± 0.04	5.64 ± 0.03	150.8 ± 7.2	0.62 ± 0.03	35.77 ± 0.66	80.60
SPED	0.02	37.81 ± 1.97	279.6 ± 1.8	0.82 ± 0.07	1.05 ± 0.06	660.9 ± 5.6	0.67 ± 0.02	15.12 ± 0.72	54.10
	0.04	70.23 ± 2.66	198.5 ± 1.9	0.85 ± 0.03	3.09 ± 0.02	435.7 ± 4.8	0.66 ± 0.04	22.94 ± 0.35	69.75
	0.08	81.23 ± 2.74	109.1 ± 2.0	0.87 ± 0.06	6.07 ± 0.05	278.5 ± 3.9	0.59 ± 0.03	29.14 ± 0.46	76.18
	0.12	96.16 ± 2.83	65.7 ± 1.3	0.88 ± 0.05	9.01 ± 0.06	99.4 ± 4.3	0.65 ± 0.06	48.85 ± 0.58	85.79

3.2. Adsorption Isotherm and Thermodynamic Calculation

Information on the interaction between the perimidine derivative and the metal surface can be obtained from the adsorption isotherm. The surface coverage (θ) can be obtained from PDP measurements. The PDP results of PMD and SPMD were plotted as C/θ versus C to obtain two adsorption curves that almost showed a straight line and a linear correlation coefficient (R^2) > 0.99 (Figure 5). It can be seen that the adsorption of perimidine derivatives on the iron obeys the Langmuir isotherm, and the formula is as follows [37]:

$$\frac{C}{\theta} = K_{ads}^{-1} + C \quad (10)$$

where C is the concentration of the perimidine derivative, K_{ads} is the equilibrium constant of adsorption, and θ is the surface coverage rate of steel [38]. In addition, the standard adsorption free energy (ΔG^0_{ads}) can be defined as follows (11):

$$\Delta G^0_{ads} = -RT \ln(55.5 K_{ads}) \quad (11)$$

where R is the molar gas constant, which is 8.314 J/(mol·k), and T is the thermodynamic temperature [39].

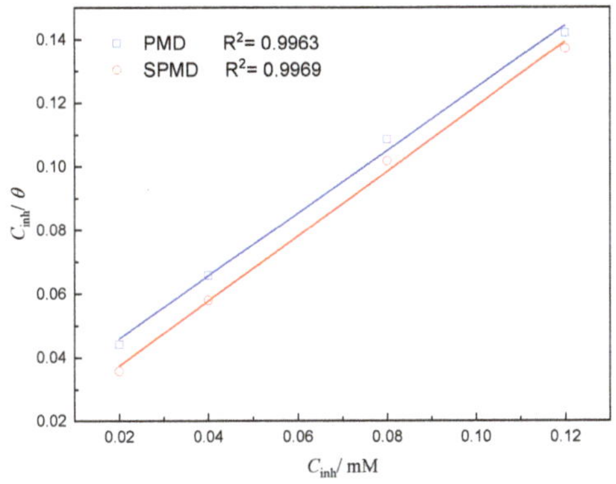

Figure 5. Curve fitting of corrosion data for reinforced steel in SCP solution with different concentrations of perimidine derivatives to Langmuir adsorption isotherm at 25 °C.

Table 3 presents the thermodynamic parameters from the Langmuir adsorption isotherm. The ΔG^0_{ads} values for PMD and SPMD are −36.11 and −37.18 kJ/mol, respectively, suggesting that the perimidine derivative molecules can be spontaneously adsorbed on iron. Generally, the absolute ΔG^0_{ads} value is between 20 and 40 kJ·mol^{-1}, indicating that the perimidine derivatives undergo both physical and chemical adsorption through electrostatic action on the surface of the metal [40,41]. Furthermore, the larger the adsorption equilibrium constant of SPMD means that there is a stronger force between SPMD and the metal surface than PMD.

Table 3. Thermodynamic and equilibrium adsorption parameters for inhibitor adsorption behavior on iron in SCP solution.

Inhibitors	K_{ads} (L·mol^{-1})	ΔG^0_{ads} (kJ/mol)
PMD	38,240.9	−36.11
SPMD	58,858.2	−37.18

3.3. Surface Morphology

In order to further verify the corrosion inhibition effect of PMD and SPMD on reinforced steel, SEM was used to analyze the morphology of the steel before and after immersion. Figure 6 displays the SEM pictures of specimens immersed in SCP in the absence and presence of 0.12 mM perimidine derivatives after 3 days: (a) before submersion, (b) after submersion in the SCP solution without an inhibitor, (c) after submersion in SCP with 0.12 mM PMD, (d) after submersion in SCP with 0.12 mM SPMD. The reinforced steel surface is smooth without holes, and there are distinct abrading stripes (Figure 6a). Due to the corrosion effect of chlorine ions on reinforced steel, the surface has obvious pitting holes and has a seriously damaged morphology in the uninhibited SCP solution (Figure 6b). After immersion in SCP with an inhibitor, the corrosion of the iron surface is significantly improved (Figure 6c,d), and the abrading stripes can still be observed. Because the perimidine derivatives adsorb on iron to form a film to protect the steel, erosion by chloride ions is prevented. Furthermore, the clearer the polishing stripe of steel submersed in SCP with SPMD, the fewer the pitting holes in the steel. The results indicate that SPMD is more effective than PMD in the concentration range of this study.

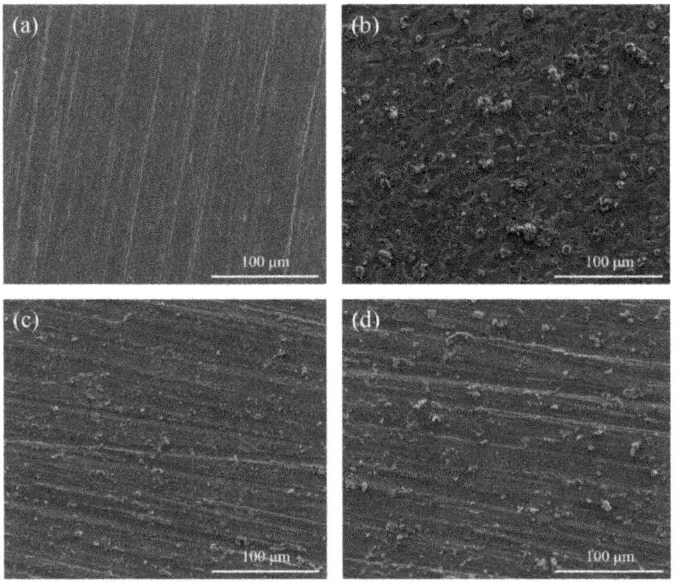

Figure 6. SEM of the specimen after submersion in SCP solution with inhibitors. (**a**) Clean specimen; (**b**) reinforced steel in SCP; (**c**) reinforced steel in SCP with 0.12 mM PMD; (**d**) reinforced steel in SCP with 0.12 mM SPMD.

3.4. Contact Angle and Surface Free Energy

The adsorption of perimidine derivatives can be tested by performing a contact angle experiment. The surface free energy value calculation formula is as follows [42,43]:

$$W_A = \gamma_{lv}(1 + \cos\alpha) \quad (12)$$

$$W_A = 2(\gamma_{lv}\gamma_{sv})^{1/2}\exp\left[-\beta(\gamma_{lv} - \gamma_{sv})^2\right] \quad (13)$$

where γ_{lv} is the surface tension of water, γ_{sv} is the surface free energy of the iron specimen, α is the contact angle, W_A is the work of adhesion, and β is a constant value, which is 0.0001247 $(mJ/m^2)^{-2}$ [44]. The contact angle images for reinforced steel in the SCP solution in the absence and presence of 0.12 mM PED and HPED are shown in Figure 7.

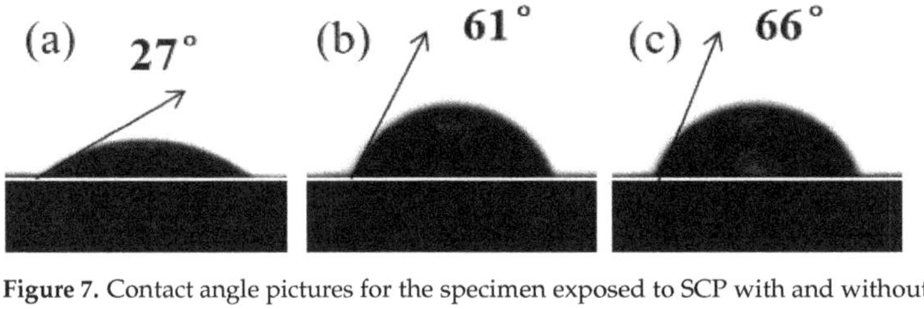

Figure 7. Contact angle pictures for the specimen exposed to SCP with and without 0.12 mM PMD and SPMD: (**a**) SCP; (**b**) SCP + 0.12 mM PMD; (**c**) SCP + 0.12 mM SPMD.

From Table 4, it is evident that the order of the contact angles of HRB400 reinforced steel in different solutions is α_{SCP} (27°) < α_{PED} (61°) < α_{HPED} (66°). The low α value is due to the presence of oxides and corrosives on iron that caused water droplets to spread on the sample surface [45]. The α value of SPMD is higher than that of PMD, indicating that SPMD has fewer corrosion products on iron. In addition, with the addition of perimidine derivatives, the γ_{sv} value of SPMD decreases, and the γ_{sv} value of SPMD is lower than that of PMD. In other words, the higher the γ_{sv} value, the more active the metal surface, and the more likely chloride ions are to corrode the metal. Therefore, these results suggest that SPMD has higher inhibition efficiency according to α and γ_{sv}.

Table 4. Obtained α and γ_{sv} values for samples exposed to SCP without and with 0.12 mM PMD and SPMD.

	Blank	PMD	SPMD
α (deg)	27	61	66
γ_{sv} (mJ/m^2)	65.15	46.62	43.59

3.5. Quantum Chemical Calculations

Figure 8 shows the PMD and SPMD molecular structure optimization, ESP and the frontier molecular orbitals (HOMO and LUMO). When the molecular structure is optimized, the frequency calculation is positive, and the molecules have reached the ground state, indicating that the potential surface was at the minimum value. The study of the lowest unoccupied orbital (LUMO) and the highest occupied orbital (HOMO) is an indispensable method in analyzing the adsorption effect of perimidine derivatives [46,47]. For molecular structure optimization, the fusion of large bonds with the N-heterocyclic ring makes PMD and SPMD appear as planar structures. The HOMO distribution of PMD and SPMD is relatively uniform in the whole molecular structure, which indicates that both the large bond and the N-heterocyclic ring can provide electrons to metal atoms. However, the LUMO of PMD is obviously concentrated in N-heterocyclic fragments and one of the benzene rings, and the distribution is not uniform, which also indicates that these fragments play a more important role in receiving electrons. While the LUMO distribution of SPMD is relatively uniform, each fragment has the same ability to receive electrons from metal.

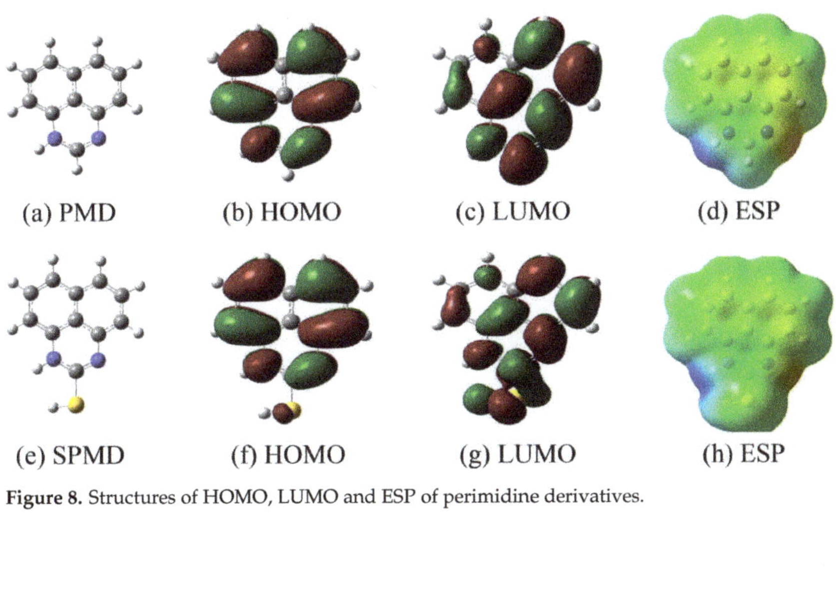

Figure 8. Structures of HOMO, LUMO and ESP of perimidine derivatives.

In order to quantitatively measure the reactivity of PMD and SPMD and the ability to donate/receive electrons, the global quantization parameters are presented in Table 5. The highest occupied orbital energy (E_{HOMO}) reflects the ability to donate electrons. The greater the E_{HOMO} value, the stronger the electron-donating ability of the molecule. The lowest unoccupied orbital energy (E_{LUMO}) reflects its ability to accept electrons. Furthermore, the difference between the two is the energy gap (ΔE); the smaller its value, the stronger the adsorption of the molecules on iron, and the better effect [48]. From Table 5, the energy gap (ΔE) of SPMD is smaller than that of PMD, indicating that SPMD has a stronger electron-donating ability [49].

Table 5. Quantum chemical parameters of perimidine derivatives.

Compounds	E_{HOMO}/ev	E_{LUMO}/ev	ΔE/ev	μ/(Debye)	χ	η	ΔN
PMD	−4.88	−0.64	4.24	3.02	2.758	2.12	1.00
SPMD	−4.96	−0.85	4.11	3.37	2.906	2.05	1.01

The 2-position sulfhydryl group of SPMD tends to bond with the empty *d* orbital of the iron atom [13], and its strong electron-accepting ability means that SPMD has a better inhibitory effect [50]. In addition, electronegativity is related to chemical potential. A higher electronegativity means a greater effect. Table 5 shows that $\chi_{SPMD} > \chi_{PMD}$, indicating that the inhibition efficiency of SPMD is better than that of PMD, which is also consistent with the experimental results. The transferred electrons (ΔN) can be calculated by Equation (6), and $\Delta N > 0$ indicates electron transfer to the metal atoms [18,51]. The results reveal that the perimidine derivatives provide electrons to iron to form coordination bonds and are more strongly adsorbed on the metal.

3.6. Mechanism of Adsorption

Figure 9 clearly describes the adsorption mechanism of perimidine derivatives on HRB400 steel. The perimidine derivatives are effectively adsorbed on iron. Through physical, chemical and re-chemisorption on iron, it can inhibit corrosion by Cl^- on steel in an alkaline medium [52]. Nitrogen atoms in perimidine derivatives are physically adsorbed by protonation with negatively charged chloride ions on the steel. In addition, the lone pair of electrons in the heteroatoms are combined with the empty *d* orbital of iron atoms and adsorbed on the iron through chemical action [53]. For example, the lone pair of electrons of nitrogen and sulfur atoms in the SPMD molecule can coordinate with iron atoms to form bonds and adsorb strongly on the metal surface, preventing charge transfer in the corrosion process and enhancing the stability of the protective film. Furthermore, retro-donation occurs, in which the filled iron orbitals share d-electron pairs with the vacant anti-bonding π orbitals in the phenyl rings. The delocalized π-electron pairs transfer to the vacant 3*d* orbitals of iron to promote the chemical action, which is re-chemisorption. It can be concluded that perimidine derivatives isolated iron from the free Cl^- by forming a protective layer.

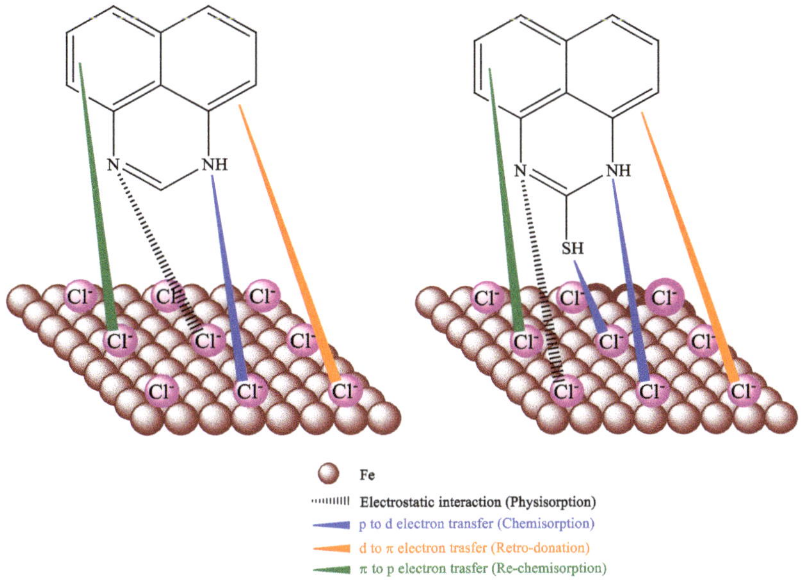

Figure 9. Inhibition and adsorption mechanisms of PMD and SPMD.

4. Conclusions

(1) PMD and SPMD are perimidine derivatives, which can inhibit the corrosion of reinforced steels in a simulated concrete pore solution. With the increase in the concentration of the inhibitor, the inhibition efficiency is gradually improved. The perimidine derivatives can effectively inhibit the corrosion of chloride ions. Moreover, SPMD has a better inhibition efficiency in the concentration range of this study.

(2) The adsorption types of perimidine derivatives on reinforced steel obey Langmuir adsorption isotherms. The standard adsorption free energy values for PMD and SPMD are −36.11 and −37.18 kJ/mol, respectively. The inhibitor molecules can spontaneously adsorb on iron to form a protective film, and the interaction is stronger between SPMD and the iron surface than PMD.

(3) The various chemical parameters of PMD and SPMD can be calculated through quantum chemical calculations. The energy gap of SPMD is smaller than that of PED, and SPMD has a stronger ability to accept and donate electrons. In addition, the electronegativity of SPMD is greater than that of PMD. Therefore, SPMD exhibits a higher inhibition efficiency.

Author Contributions: Conceptualization, S.M. and Q.M.; methodology, J.C.; software, H.Z.; formal analysis, Q.M.; resources, J.L.; data curation, K.L.; writing—review and editing, Q.M.; project administration, J.H.; funding acquisition, S.M. All authors have read and agreed to the published version of the manuscript.

Funding: This work was supported by the National Key R & D Program of China (No. 2021YFF0500803), the National Natural Science Foundation of China (Grant Nos. 51908254 and 52078240) and the National Science Foundation for Outstanding Youth project of China (Grant No. 51825203).

Institutional Review Board Statement: Not applicable.

Informed Consent Statement: Not applicable.

Data Availability Statement: Not applicable.

Conflicts of Interest: The authors declare no conflict of interest.

References

1. Zhi, F.; Jiang, L.; Jin, M.; Xu, P.; Xiao, B.; Jiang, Q.; Chen, L.; Gu, Y. Inhibition effect and mechanism of polyacrylamide for steel corrosion in simulated concrete pore solution. *Constr. Build. Mater.* **2020**, *259*, 120425. [CrossRef]
2. Wang, H.; Zhang, A.; Zhang, L.; Liu, J.; Han, Y.; Shu, H.; Wang, J. Study on the influence of compound rust inhibitor on corro-sion of steel bars in chloride concrete by electrical parameters. *Constr. Build. Mater.* **2020**, *262*, 120763–120777. [CrossRef]
3. Verma, C.; Quraishi, M.; Ebenso, E.E. Quinoline and its derivatives as corrosion inhibitors: A review. *Surf. Interfaces* **2020**, *21*, 100634. [CrossRef]
4. Thompson, N.G.; Yunovich, M.; Dunmire, D. Cost of corrosion and corrosion maintenance strategies. *Corros. Rev.* **2007**, *25*, 247–262. [CrossRef]
5. Fajardo, G.; Escadeillas, G.; Arliguie, G. Electrochemical chloride extraction (ECE) from steel-reinforced concrete specimens contaminated by "artificial" sea-water. *Corros. Sci.* **2006**, *48*, 110–125. [CrossRef]
6. Bastidas, D.; Criado, M.; La Iglesia, V.; Fajardo, S.; La Iglesia, A. Comparative study of three sodium phosphates as corrosion inhibitors for steel reinforcements. *Cem. Concr. Compos.* **2013**, *43*, 31–38. [CrossRef]
7. Pan, C.; Chen, N.; He, J.; Liu, S.; Chen, K.; Wang, P.; Xu, P. Effects of corrosion inhibitor and functional components on the elec-trochemical and mechanical properties of concrete subject to chloride environment. *Constr. Build. Mater.* **2020**, *260*, 119724–119737. [CrossRef]
8. Sánchez, M.; Alonso, M. Electrochemical chloride removal in reinforced concrete structures: Improvement of effectiveness by simultaneous migration of calcium nitrite. *Constr. Build. Mater.* **2011**, *25*, 873–878. [CrossRef]
9. Valcarce, M.B.; Vazquez, M. Carbon steel passivity examined in alkaline solutions: The effect of chloride and nitrite ions. *Electrochim. Acta* **2008**, *53*, 5007–5015. [CrossRef]
10. Shukla, S.K.; Quraishi, M.; Prakash, R. A self-doped conducting polymer "polyanthranilic acid": An efficient corrosion inhibitor for mild steel in acidic solution. *Corros. Sci.* **2008**, *50*, 2867–2872. [CrossRef]
11. Ramezanzadeh, M.; Bahlakeh, G.; Ramezanzadeh, B.; Sanaei, Z. Adsorption mechanism and synergistic corrosion-inhibiting effect between the green Nettle leaves extract and Zn2+ cations on carbon steel. *J. Ind. Eng. Chem.* **2019**, *77*, 323–343. [CrossRef]
12. Feng, L.; Yang, H.; Wang, F. Experimental and theoretical studies for corrosion inhibition of carbon steel by imidazoline deriv-ative in 5% NaCl saturated Ca(OH)2 solution. *Electrochim. Acta* **2011**, *58*, 427–436. [CrossRef]
13. He, X.; Mao, J.; Ma, Q.; Tang, Y. Corrosion inhibition of perimidine derivatives for mild steel in acidic media: Electrochemical and computational studies. *J. Mol. Liq.* **2018**, *269*, 260–268. [CrossRef]
14. Herbert, J.; Woodgate, P.; Denny, W. Potential antitumor agents. 53. Synthesis, DNA binding properties, and biological activity of perimidines designed as minimal DNA-intercalating agents. *J. Med. Chem.* **1987**, *30*, 2081–2086. [CrossRef] [PubMed]
15. Gong, J.; Sturla, S.J. A Synthetic Nucleoside Probe that Discerns a DNA Adduct from Unmodified DNA. *J. Am. Chem. Soc.* **2007**, *129*, 4882–4883. [CrossRef]
16. Xu, P.; Zhou, J.; Li, G.; Wang, P.; Wang, P.; Li, F.; Zhang, B.; Chi, H. Corrosion inhibition efficiency of compound nitrite with D-sodium gluconate on carbon steel in simulated concrete pore solution. *Constr. Build. Mater.* **2021**, *288*, 123101. [CrossRef]
17. Sundaram, R.G.; Vengatesh, G.; Sundaravadivelu, M. Surface morphological and quantum chemical studies of some expired drug molecules as potential corrosion inhibitors for mild steel in chloride medium. *Surf. Interfaces* **2020**, *22*, 100841. [CrossRef]
18. Saha, S.; Banerjee, P. A theoretical approach to understand the inhibition mechanism of steel corrosion with two aminoben-zonitrile inhibitors. *RSC Adv.* **2015**, *5*, 71120–71130. [CrossRef]
19. Guo, L.; Safi, Z.; Shi, W.; Tüzün, B.; Altunay, N.; Kaya, C. Anticorrosive effects of some thiophene derivatives against the corrosion of iron: A computational study. *Front. Chem.* **2018**, *6*, 155–167. [CrossRef]
20. Dohare, P.; Ansari, K.; Quraishi, M.; Obot, I. Pyranpyrazole derivatives as novel corrosion inhibitors for mild steel useful for industrial pickling process: Experimental and Quantum Chemical study. *J. Ind. Eng. Chem.* **2017**, *52*, 197–210. [CrossRef]
21. Haque, J.; Ansari, K.; Srivastava, V.; Quraishi, M.; Obot, I. Pyrimidine derivatives as novel acidizing corrosion inhibitors for N80 steel useful for petroleum industry: A combined experimental and theoretical approach. *J. Ind. Eng. Chem.* **2017**, *49*, 176–188. [CrossRef]
22. Jiang, S.; Jiang, L.; Wang, Z.; Jin, M.; Bai, S.; Song, S.; Yan, X. Deoxyribonucleic acid as an inhibitor for chloride-induced corro-sion of reinforcing steel in simulated concrete pore solutions. *Constr. Build. Mater.* **2017**, *150*, 238–247. [CrossRef]
23. Wang, Y.; Zuo, Y.; Tang, Y. Inhibition effect and mechanism of sodium oleate on passivation and pitting corrosion of steel in simulated concrete pore solution. *Constr. Build. Mater.* **2018**, *167*, 197–204. [CrossRef]
24. Zhou, X.; Yang, H.; Wang, F. Investigation on the inhibition behavior of a pentaerythritol glycoside for carbon steel in 3.5% NaCl saturated Ca(OH)2 solution. *Corros. Sci.* **2012**, *54*, 193–200. [CrossRef]
25. Sanaei, Z.; Ramezanzadeh, M.; Bahlakeh, G.; Ramezanzadeh, B. Use of Rosa canina fruit extract as a green corrosion inhibitor for mild steel in 1 M HCl solution: A complementary experimental, molecular dynamics and quantum mechanics investigation. *J. Ind. Eng. Chem.* **2019**, *69*, 18–31. [CrossRef]

26. Azeez, F.A.; Al-Rashed, O.A.; Nazeer, A.A. Controlling of mild-steel corrosion in acidic solution using environmentally friendly ionic liquid inhibitors: Effect of alkyl chain. *J. Mol. Liq.* **2018**, *265*, 654–663. [CrossRef]
27. Zhao, Y.; Pan, T.; Yu, X.; Chen, D. Corrosion inhibition efficiency of triethanolammonium dodecylbenzene sulfonate on Q235 carbon steel in simulated concrete pore solution. *Corros. Sci.* **2019**, *158*, 108097–108109. [CrossRef]
28. Yamaguchi, M.; Nishihara, H.; Aramaki, K. The inhibition of passive film breakdown on iron in a borate buffer solution con-taining chloride ions by organic anion inhibitors. *Corros. Sci.* **1994**, *36*, 241–258. [CrossRef]
29. Valek, L.; Martinez, S.; Mikulić, D.; Brnardić, I. The inhibition activity of ascorbic acid towards corrosion of steel in alkaline media containing chloride ions. *Corros. Sci.* **2008**, *50*, 2705–2709. [CrossRef]
30. Yamaguchi, M.; Nishihara, H.; Aramaki, K. The inhibition of pit growth on an iron surface in a borate buffer solution containing chloride ion by inhibitors classified as soft bases in the HSAB principle. *Corros. Sci.* **1995**, *37*, 571–585. [CrossRef]
31. Yang, D.; Zhang, M.; Zheng, J.; Castaneda, H. Corrosion inhibition of mild steel by an imidazolium ionic liquid compound: The effect of pH and surface pre-corrosion. *RSC Adv.* **2015**, *5*, 95160–95170. [CrossRef]
32. Hu, J.; Koleva, D.; Petrov, P.; Breugel, K. Polymeric vesicles for corrosion control in reinforced mortar: Electrochemical behav-ior, steel surface analysis and bulk matrix properties. *Corros. Sci.* **2012**, *65*, 414–430. [CrossRef]
33. John, D.G.; Searson, P.C.; Dawson, J.L. Use of AC impedance technique in studies on steel in concrete in immersed conditions. *Br. Corros. J.* **1981**, *16*, 102–106. [CrossRef]
34. Ansari, K.; Quraishi, M. Bis-Schiff bases of isatin as new and environmentally benign corrosion inhibitor for mild steel. *J. Ind. Eng. Chem.* **2014**, *20*, 2819–2829. [CrossRef]
35. Al-Mehthel, M.; Al-Dulaijan, S.; Al-Idi, S.H.; Shameem, M.; Ali, M.; Maslehuddin, M. Performance of generic and proprietary corrosion inhibitors in chloride-contaminated silica fume cement concrete. *Constr. Build. Mater.* **2009**, *23*, 1768–1774. [CrossRef]
36. Gunay, H.B.; Ghods, P.; Isgor, O.B.; Carpenter, G.J.; Wu, X. Characterization of atomic structure of oxide films on carbon steel in simulated concrete pore solutions using EELS. *Appl. Surf. Sci.* **2013**, *274*, 195–202. [CrossRef]
37. Lebrini, M.; Suedile, F.; Salvin, P.; Roos, C.; Zarrouk, A.; Jama, C.; Bentiss, F. Bagassa guianensis ethanol extract used as sustainable eco-friendly inhibitor for zinc corrosion in 3% NaCl: Electrochemical and XPS studie. *Surf. Interfaces* **2020**, *20*, 100588–100603. [CrossRef]
38. Hemapriya, V.; Prabakaran, M.; Parameswari, K.; Chitra, S.; Kim, S.; Chung, I. Dry and wet lab analysis on benzofused hetero-cyclic compounds as effective corrosion inhibitors for mild steel in acidic medium. *J. Ind. Eng. Chem.* **2016**, *40*, 106–117. [CrossRef]
39. Ma, Q.; Qi, S.; He, X.; Tang, Y.; Lu, G. 1,2,3-Triazole derivatives as corrosion inhibitors for mild steel in acidic medium: Experimental and computational chemistry studies. *Corros. Sci.* **2017**, *129*, 91–101. [CrossRef]
40. Herrag, L.; Hammouti, B.; Elkadiri, S.; Aouniti, A.; Jama, C.; Vezin, H.; Bentiss, F. Adsorption properties and inhibition of mild steel corrosion in hydrochloric solution by some newly synthesized diamine derivatives: Experimental and theoretical investigations. *Corros. Sci.* **2010**, *52*, 3042–3051. [CrossRef]
41. Zhang, Z.; Ba, H.; Wu, Z. Sustainable corrosion inhibitor for steel in simulated concrete pore solution by maize gluten meal extract: Electrochemical and adsorption behavior studies. *Constr. Build. Mater.* **2019**, *227*, 117080. [CrossRef]
42. Alinejad, S.; Naderi, R.; Mahdavian, M. Effect of inhibition synergism of zinc chloride and 2-mercaptobenzoxzole on protec-tive performance of an ecofriendly silane coating on mild steel. *J. Ind. Eng. Chem.* **2017**, *48*, 88–98. [CrossRef]
43. Li, D.; Neumann, A.W. A reformulation of the equation of state for interfacial tensions. *J. Colloid Interface Sci.* **1990**, *137*, 304–307. [CrossRef]
44. Guo, Y.; Xu, B.; Liu, Y.; Yang, W.; Yin, X.; Chen, Y.; Le, J.; Chen, Z. Corrosion inhibition properties of two imidazolium ionic liq-uids with hydrophilic tetrafluoroborate and hydrophobic hexafluorophosphate anions in acid medium. *J. Ind. Eng. Chem.* **2017**, *56*, 234–247. [CrossRef]
45. Pandarinathan, V.; Lepková, K.; Bailey, S.I.; Becker, T.; Gubner, R. Adsorption of corrosion inhibitor 1-Dodecylpyridinium chloride on carbon steel studied by *in Situ* AFM and electrochemical methods. *Ind. Eng. Chem. Res.* **2014**, *53*, 5858–5865. [CrossRef]
46. Hejazi, S.; Mohajernia, S.; Moayed, M.; Davoodi, A.; Rahimizadeh, M.; Momeni, M.; Eslami, A.; Shiri, A.; Kosari, A. Electrochemical and quantum chemical study of Thiazolo-pyrimidine derivatives as corrosion inhibitors on mild steel in 1 M H_2SO_4. *J. Ind. Eng. Chem.* **2015**, *25*, 112–121. [CrossRef]
47. Alrefaee, S.H.; Rhee, K.Y.; Verma, C.; Quraishi, M.; Ebenso, E.E. Challenges and advantages of using plant extract as inhibitors in modern corrosion inhibition systems: Recent advancements. *J. Mol. Liq.* **2020**, *321*, 114666. [CrossRef]
48. Obi-Egbedi, N.; Obot, I. Inhibitive properties, thermodynamic and quantum chemical studies of alloxazine on mild steel corro-sion in H_2SO_4. *Corros. Sci.* **2011**, *53*, 263–275. [CrossRef]
49. Eddy, N.; Awe, F.; Gimba, C.; Ibisi, N.; Ebenso, E. Experimental and Computational Chemistry Simulation Studies on the inhi-bition potentials of some amino acids for the corrosion of mild steel in 0.1 M. *Int. J. Electrochem. Sci.* **2011**, *6*, 931–957.
50. Machnikova, E.; Whitmire, K.H.; Hackerman, N. Corrosion inhibition of carbon steel in hydrochloric acid by furan derivatives. *Electrochim. Acta* **2008**, *53*, 6024–6032. [CrossRef]

51. Masoud, M.; Awad, M.; Shaker, M.; El-Tahawy, M. The role of structural chemistry in the inhibitive performance of some aminopyrimidines on the corrosion of steel. *Corros. Sci.* **2010**, *52*, 2387–2396. [CrossRef]
52. Berdimurodov, E.; Kholikovb, A.; Akbarov, K.; Guo, L. Inhibition properties of 4,5-dihydroxy-4,5-di-p-tolylimidazolidine-2-thione for use on carbon steel in an aggressive alkaline medium with chloride ions: Thermodynamic, electrochemical, surface and theoretical analyses. *J. Mol. Liq.* **2021**, *327*, 114813–114830. [CrossRef]
53. Verma, C.; Ebenso, E.; Quraishi, M.; Rhee, K. Phthalocyanine, naphthalocyanine and their derivatives as Corrosion Inhibi-tors: A review. *J. Mol. Liq.* **2021**, *334*, 116441–116449. [CrossRef]

coatings

Article

Effect of Phaeodactylum Tricornutum in Seawater on the Hydration of Blended Cement Pastes

Junjie Wang [1], Lei Xu [1], Jiatong Guo [1], Yimu Jiang [1], Hang He [2], Yuli Wang [2,*], Weiqi Fu [3,4,5,†], Yi Zhu [6], Zhe Ye [1], Pukang He [1] and Yi Zhang [1]

1 Department of Civil Engineering, Tsinghua University, Beijing 100084, China
2 School of Materials Science and Engineering, Henan Polytechnic University, Jiaozuo 454003, China
3 Department of Marine Science, Ocean College, Zhejiang University, Zhoushan 316021, China
4 Donghai Laboratory, Zhoushan 316021, China
5 Center for Systems Biology and Faculty of Industrial Engineering, Mechanical Engineering and Computer Science, School of Engineering and Natural Sciences, University of Iceland, 101 Reykjavik, Iceland
6 Shenzhen Yichuang International Design Co., Ltd., Shenzhen 518000, China
* Correspondence: wangyuli@hpu.edu.cn
† Senior author.

Abstract: Seawater can be used as mixing water for concrete with no steel reinforcement in some areas with difficult access to fresh water. Diatoms such as Phaeodactylum tricornutum are among the most abundant micro-organisms living in seawater, and they could be unavoidable when collecting seawater. In fact, diatoms can provide bio-SiO_2 and bio-$CaCO_3$ sources, namely amorphous nano-SiO_2 and crystallised nano-$CaCO_3$, which could be beneficial to cement hydration. Thus, the effects of different Phaeodactylum tricornutum concentrations (0%, 2.5% and 5% by weight of suspension of seawater and diatoms) in seawater on cement hydration in ordinary Portland cement (OPC) mixes (100% OPC) and ground granulated blast-furnace slag (GGBS) mixes (70% OPC + 30% GGBS) were investigated through tests of compressive strength, XRD, DTG–DTA and SEM. The results show that diatoms accelerated cement hydration by providing the nucleus for C-S-H structure and contributed pozzolanic reactions by amorphous nano-SiO_2 and nano-$CaCO_3$. The accelerated cement hydration was also confirmed by the fact that more $Ca(OH)_2$ was formed in cement pastes with diatoms. However, it has also been found that diatoms decreased the compressive strength of cement pastes by leaving more weak bonds between the C-S-H structure, which was considered to be caused by the organic parts and the micron gap formed in diatoms. When comparing an OPC paste mix with 5% diatoms to a blank OPC paste, the reduction in compressive strength at 28 days can reach a maximum of 50.1%. The ability to provide bridging effects between C-S-H particles in GGBS paste was discovered to depend on the development of additional ettringite. This resulted in a 7.6% loss in compressive strength after 28 days in a GGBS paste with 5% diatoms.

Keywords: diatom; cement hydration; amorphous nano-SiO_2; strength; microstructure

1. Introduction

In some areas, such as the UAE in the Middle East, with large areas of desert but adjacent to the sea, the use of seawater in plain concrete is unavoidable [1–11]. Recent studies [1,2] have reported that the application of seawater as mixing water in cementitious materials was beneficial for concrete strength development because of the accelerated hydration of cement caused by the salts in seawater [12], and an increase at even above 50% on the compressive strength of ordinary Portland cement (OPC) paste was reported by Wang et al. [1,2]. However, some previous studies showed a reduction in compressive strength when seawater was used as mixing water [12,13]. These conflicts could be caused by the different seawaters used, such as seawater that was filtered through 5 μm by Wang et al. [1,2], but some researchers [12,14] may use unfiltered seawater because filtration steps take more time and thus increase costs. Impurities such as micro-organisms in seawater could decrease the strength of cementitious materials.

Many micro-organisms, such as diatoms, exist abundantly in seawater and cannot be avoided when pumping water from the sea directly. Locally accumulated diatoms in seawater could exceed 5% by weight of seawater. Thus, the effects of impurities such as micro-organisms on the physical properties of cementitious materials need to be investigated. For simplicity, one of the most abundant micro-organisms in seawater, the diatom, was considered in this study. It is known that diatoms are rich in amorphous nano-SiO_2 [15–20], which could contribute to the hydration of cement. Nano-SiO_2 particles [21–27] could increase the compressive strength of concrete by increasing compactness and providing the nucleus for C-S-H. However, the use of diatoms in seawater also has drawbacks, such as high water demand [28] and a sudden decrease in compressive strength with an increase in dosages [29]. Thus, the substitution quantities of diatoms are generally limited to 10% without compromising strength [28,30]. To the best of the authors' knowledge, no research has ever been conducted on how diatoms affect the hydration of blended cement paste. Thus, the effect of different concentrations of diatoms in seawater (0%, 2.5% and 5% by weight of suspension with seawater and diatoms) on the compressive strength, hydration products and microstructure of OPC and blended cement pastes was investigated. Two types of cement (100% OPC, 70% OPC + 30% GGBS) were considered, to obtain the effect of diatoms in different types of cementitious materials. In this study, the high concentrations of diatoms (2.5% and 5% by weight of seawater and diatoms) were used in order to obtain an obvious effect of the diatoms. Mechanisms of diatoms on cement hydration were analysed through XRD, DTG-DTA and SEM results.

2. Materials and Methods

2.1. Materials

2.1.1. Diatoms

Figure 1a shows the common diatoms contained in seawater (used as mixing water) for the cementitious materials in this study. The name of the diatom is Phaeodactylum tricornutum (strain CCAP 1055/1) (CCAP: the culture collection of algae and protozoa). It can be seen that there are two shapes of this diatom, which is because of the different Si contents in their bodies [15,16]. Diatoms were ground into powder using a pestle and mortar. The powder was then passed through a sieve of 75 µm. XRD patterns of diatoms were collected using a Bruker D8 powder diffractometer (Bruker, Karlsruhe, Germany) with scanning angles (2θ) ranging from 5° to 95° at a rate of 0.3 s/step and 0.02°/step using a CuKα radiation source (λ = 1.5418 Å) at 40 kV and 40 mA. Figure 1b shows the XRD patterns of a diatom powder sample, and it can be seen that there are a large number of amorphous nano-SiO_2 particles at about 20–25 degrees and obvious peaks of well-crystallised nano-$CaCO_3$ and nano-$Mg(OH)_2$. Figure 1c shows the micromorphology of a raw diatom under SEM, which exhibited a rougher and more irregular structure. The diatom powder was gold-coated and then observed under high-vacuum condition using a JSM7401 field emission scanning electron microscope (JEOL, Eching b., München, Germany).

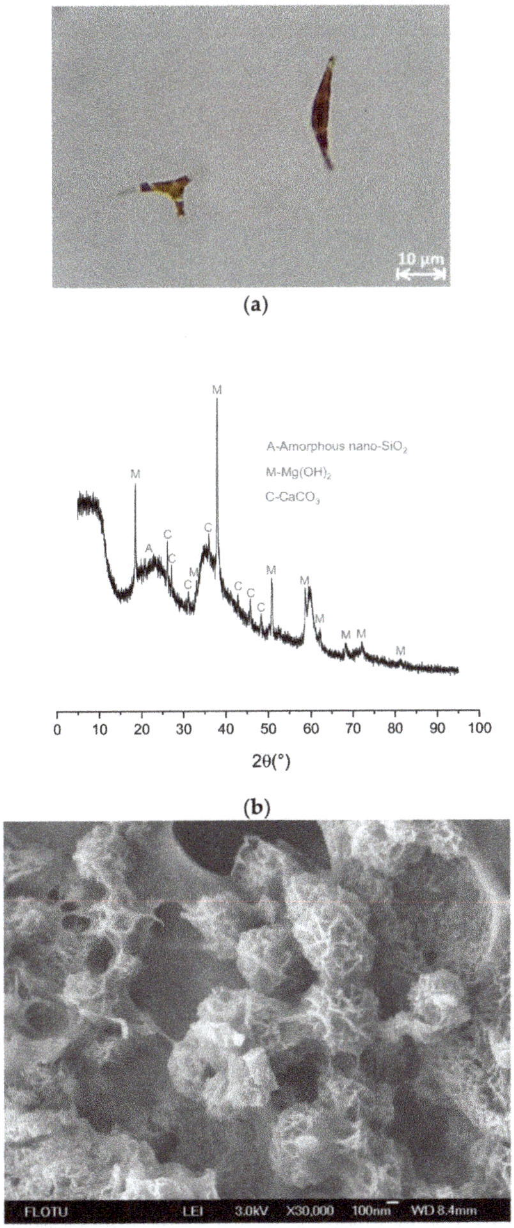

(c)

Figure 1. Characterisation of diatoms: (**a**) Observation of raw diatoms (P. tricornutum) under optical microscopy; XRD result (**b**) and SEM image (**c**) of diatom powder.

2.1.2. OPC and GGBS

The chemical compositions of OPC (type 1, conforming to ASTM C150M) and GGBS from XRF are shown in Table 1. The results show that GGBS has a much higher content of aluminium phases (Al_2O_3) than OPC and a lower content of CaO.

Table 1. Chemical composition of OPC and GGBS (%).

Composition	CaO	SiO_2	Al_2O_3	Fe_2O_3	SO_3	MgO	K_2O	Na_2O	LOI
OPC	63.02	18.97	5.32	4.26	3.01	1.56	0.54	0.29	3.03
GGBS	34.65	34.33	16.68	1.73	1.79	7.27	0.33	0.25	2.97

2.1.3. Seawater

Seawater from a local beach was collected and used as mixing water for the cementitious materials in this study. Table 2 shows the ion concentration of seawater by ICP-MS [2] after filtration through 5 μm apertures.

Table 2. Ion concentration of seawater.

Ion	Cl^-	Na^+	Mg^{2+}	$SO_4{}^{2-}$	K^+	Ca^{2+}
Concentration (g/L)	19.67	11.98	2.13	2.06	0.45	0.40

2.2. *Mix Proportions and Preparation of Test Samples*

The mix design of the samples is shown in Table 3. The effect of diatoms on the hydration of cement paste was considered from three levels of diatoms (0%, 2.5% and 5.0%) and two types of cementitious materials (OPC and GGBS mixes). OPC mixes contained 100% OPC only, and GGBS mixes contained 70% OPC + 30% GGBS. The wt.% of diatoms in seawater is the dry weight of diatoms under 60 °C for 24 h to the weight of the mixed suspension of seawater and diatoms. The different concentrations of diatoms were adjusted by adding different amounts of the filtered seawater.

Table 3. Mix proportions.

Mix ID	w/b	OPC (wt.%)	GGBS (wt.%)	Diatom (wt.%)
OPC	0.4	100	0	0
OPC + 2.5% Diatom	0.4	100	0	2.5
OPC + 5% Diatom	0.4	100	0	5.0
GGBS	0.4	70	30	0
GGBS + 2.5% Diatom	0.4	70	30	2.5
GGBS + 5% Diatom	0.4	70	30	5.0

The mixing procedure was: the OPC powder with or without GGBS was firstly mixed evenly in the mixer, and then, the seawater with different concentrations of diatoms was added with a mixing speed of 133 rpm for 4 min. The fresh pastes were then transferred into 50 mm cube moulds. The moulds with pastes inside were vibrated for 10 s and then transferred into a standard curing chamber (20 °C, R.H. 97%). All samples were demoulded after 24 h and returned to the standard curing chamber until further tests.

2.3. *Test Methods*

2.3.1. Compressive Strength

After curing for 1, 7 and 28 days in the standard curing chamber, the cubic samples were taken out for compressive strength tests. The tests were conducted in a displacement-controlled mode at a speed of 0.02 mm/s with an MTS machine (MTS Criterion 43, Mechanical Testing & Simulation, Eden Prairie, MN, USA). Three samples were tested for each group.

2.3.2. XRD

The grinding steps of hardened pastes were consistent with those of diatoms. One gram of powder was used for XRD measurement each time. A Rigaku SmartLab XRD machine (Rigaku, Tokyo, Japan) was used. The working voltage was 40 kV, and the current was 150 mA. The scanning was from 5° to 70°, and the scanning rate was 10°/min.

2.3.3. DTG/DTA

The powder samples prepared using the method above were also used for DTG/DTA measurements (HCT-3, Henven Scientific Instument Factory, Beijing, China). The samples were heated from 25 to 960 °C at a rate of 10 °C/min in a N_2 environment. Samples were stabilised for 20 min in the chamber before the measurements started.

2.3.4. SEM

Selected representative cement paste samples from the inside part of specimens were firstly gold-coated and then observed under high-vacuum condition using a Merlin Compact Field Emission SEM machine (Merlin Compact, Carl Zeiss NTS GmbH, Jena, Germany). The analyses of various morphologies were conducted at 20 kV with a 10 mm working distance.

3. Results and Discussions

3.1. Compressive Strength

Figure 2 shows the results on compressive strength of all mixes after standard curing for 1, 7 and 28 days. It can be seen that the effect of diatoms on cement hydration is different in the different mixes, and it is affected by different concentrations of diatoms in seawater and different cementitious materials. At 1 day, the effect of diatom concentration on the change of compressive strength is not obvious in both OPC and GGBS mixes, and there is only a slight decrease in compressive strength with increasing diatom concentration. However, at 7 and 28 days, the effect of diatom concentration on compressive strength became significant, especially in OPC mixes. In OPC mixes, with the increasing diatom concentrations from 0% to 2.5% and 5%, the decrease in compressive strength was 24.5% and 49.1% at 7 days, and 21.0% and 50.1% at 28 days. In GGBS mixes, with the same increasing diatom concentrations from 0% to 2.5% and 5%, the decrease in compressive strength was 5.2% and 8.6% at 7 days, and only 2.2% and 7.6% at 28 days. It is interesting to see the deleterious effect of diatom on strength development was not as significant in GGBS mixes as in OPC mixes. GGBS mixes can more effectively resist the adverse effects of diatoms on the compressive strength of paste than OPC mixes. From the chemical compositions shown in Table 3, the aluminium phases in GGBS are much higher than in OPC, so the existence of amorphous nano-SiO_2 in the diatoms [17–20] could facilitate additional reaction with the aluminium phases by forming more ettringite in GGBS mixes [31–35] compared to that in OPC mixes, by accelerating the hydration process and providing nucleus sites. The related mechanisms are discussed through the XRD, DTG/DTA and SEM results.

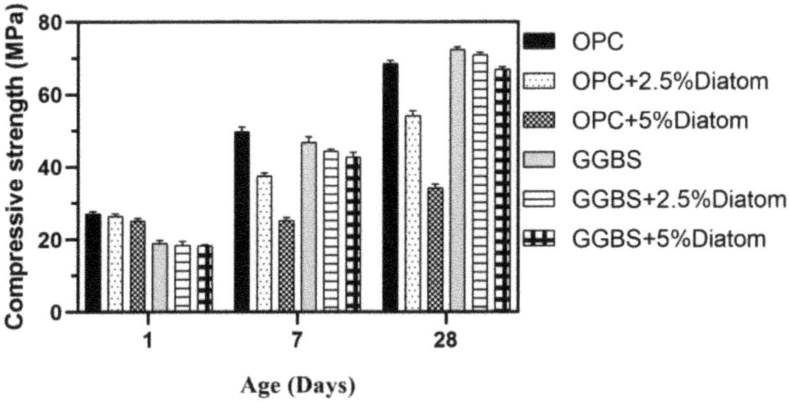

Figure 2. Compressive strength results of different samples at ages of 1, 7 and 28 days.

3.2. XRD Results

Figure 3 shows the diffractograms of the crystals identified by XRD in different samples after curing for 1 and 28 days. The relative peak heights can be related to the *Y*-axis values. It can be seen that the results in OPC mixes and GGBS mixes are different. The peak heights of $Ca(OH)_2$ in OPC mixes are generally higher than those in GGBS mixes at both 1 and 28 days, which is due to the consumption of $Ca(OH)_2$ produced by OPC hydration as an alkali activator of GGBS.

It is noted that there are obvious peaks of Friedel's salt in all OPC and GGBS mixes with or without diatoms. The existence of high amounts of chlorides provided the reaction to form Friedel's salt in cement pastes [1,2,7,8]. At 1 day (Figure 3a), the effect of different diatom concentrations on the peak heights and peak areas of $Ca(OH)_2$ was not obvious. At 28 days (Figure 3b), in both GGBS and OPC mixes, the peak heights and peak areas of $Ca(OH)_2$ increased gradually with increasing diatom concentrations from 0% to 2.5% and 5%. This could indicate accelerated hydration of C_3S by the existence of amorphous nao-SiO_2 in the diatoms, to some extent. Figure 3 also shows that the peak heights and peak areas of ettringite in GGBS mixes are higher than in OPC mixes, indicating there could be more ettringite in GGBS mixes. This is caused by the higher amounts of aluminium phases in GGBS than OPC, as shown in Table 1.

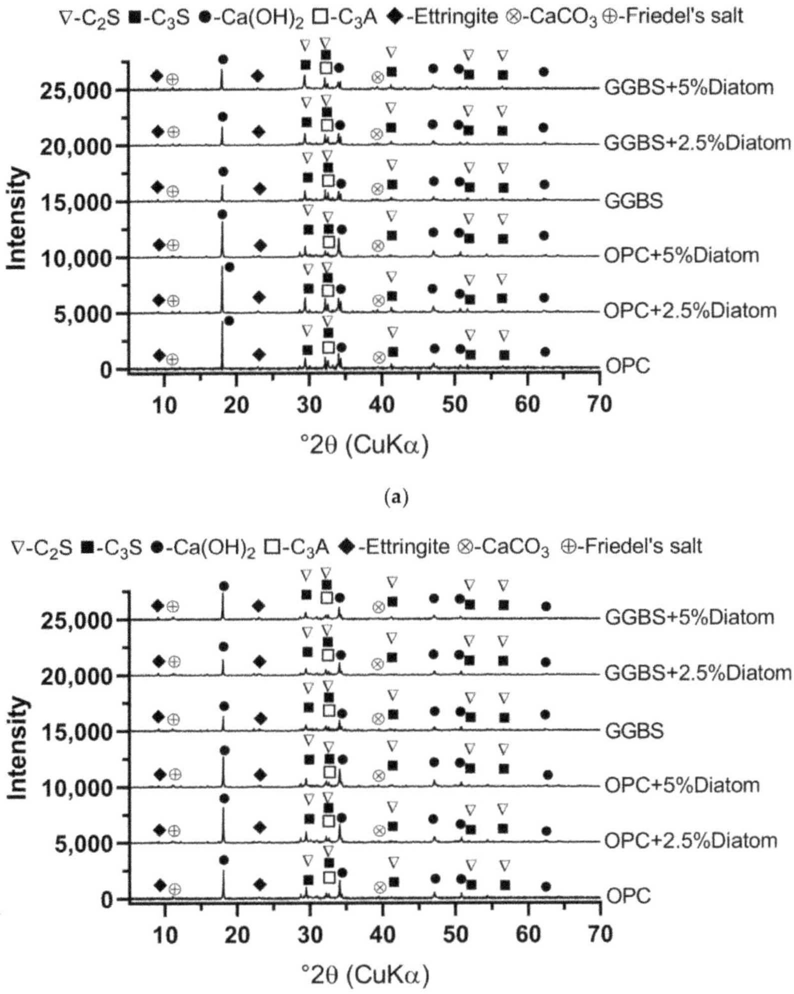

Figure 3. XRD results of different mixes: (**a**) 1 day, (**b**) 28 days.

3.3. DTG/DTA Results

Figure 4 shows the effect of different diatom concentrations on the DTG results of the OPC and GGBS mixes after curing for 1 and 28 days. It can be seen that the main peaks correspond to ettringite (80–100 °C [36,37]), Friedel's salt (120, 160 and 360 °C [1,2]), $Ca(OH)_2$ (450 °C [38,39]) and $CaCO_3$ (680–720 °C [21,38]), which agrees well with the crystals identified in XRD results (Figure 3). At 1 day, the peaks of ettringite and $Ca(OH)_2$ in OPC mixes are higher than those in GGBS mixes in Figure 3 because the GGBS has a lower content of CaO and SO_3 than OPC (from Table 1). At 28 days, it is interesting to see that the peak of ettringite in mix OPC + 5% diatom is much lower than that in mix OPC + 2.5% diatom—this could indicate that the amorphous nano-SiO_2 in diatoms may facilitate the transformation of some ettringite to Friedel's salt, which could be evidenced by the higher DTA peak (Figure 5) in the mix OPC + 5% diatom than that in the mix OPC + 2.5% diatom. The fluctuations in Figure 4 at the local range could be caused by the minimal impurities from the reaction between the trace amounts of other salts (except NaCl) and the OPC minerals or the organic parts in diatoms.

Figure 5 shows the DTA results caused by different diatom concentrations in OPC and GGBS mixes. The corresponding changes in each peak have been marked in the figure, and the differences in each peak can be related to the changes illustrated in Figure 4. The results show that the most obvious energy change occurs at the temperature range of $Ca(OH)_2$. This agrees well with the DTG results. The difference of DTA results between OPC and GGBS mixes is found to be higher than the difference caused by different diatom concentrations. The DTA peaks of $CaCO_3$ are not obvious in all mixes, indicating the energy change caused by the decomposition of $CaCO_3$ in cement pastes is not significant at this stage. Compared to the DTG results, the DTA results could not provide accurate information on the existence of $CaCO_3$ in cement pastes. Additionally, it was discovered that peaks in the DTA at the sites for ettringite and $Ca(OH)_2$ could not accurately compare the percentages of these phases.

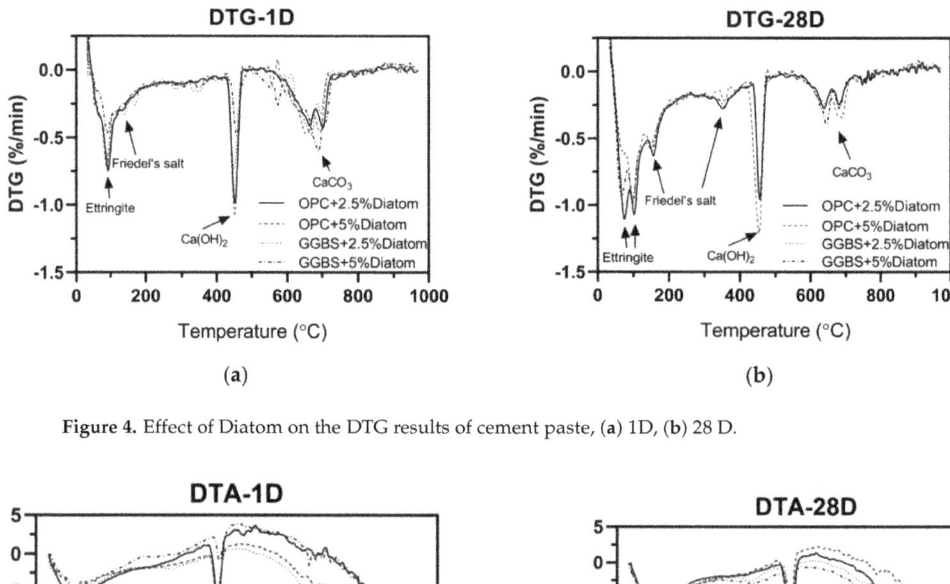

(a) (b)

Figure 4. Effect of Diatom on the DTG results of cement paste, (**a**) 1D, (**b**) 28 D.

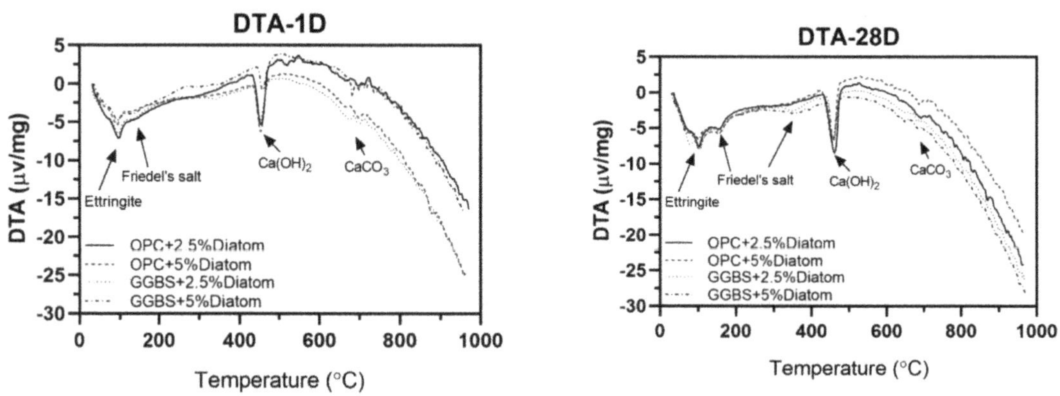

(a) (b)

Figure 5. Effect of Diatom on the DTA results of cement paste, (**a**) 1D, (**b**) 28 D.

3.4. Microstructure Analysis

Figures 6 and 7 show the microstructure under SEM observations of different OPC and GGBS mixes with 2.5% and 5% diatoms after standard curing for 1 and 28 days. The existence of ettringite (AFt), $Ca(OH)_2$ and Friedel's salt can be seen in different mixes. It is found that ettringite is more common than Friedel's salt in the four cement pastes, indicating a higher amount of ettringite was formed than Friedel's salt. This agrees well with the XRD results. It can be clearly seen that, in both OPC and GGBS mixes with 5% diatoms, there are two types of C-S-H [1,40,41], namely low-density C-S-H (LD C-S-H) and high-density C-S-H (HD C-S-H).

By comparing the mixes with 2.5% and 5% diatoms at 1 day in Figure 6, it can be found that, in both OPC and GGBS mixes, the paste with a higher concentration of diatoms (5%) had more clustered nuclear-shaped HD C-S-H. This indicates that the existence of amorphous nano-SiO_2 in diatoms [15–18] and nano-$CaCO_3$ [37,38] increased the nuclear sites for the formation of the HD C-S-H during the hydration process in cement paste within the early age (1 day). However, the higher amounts of diatoms contain a large amount of amorphous silica, which has high pozzolanic activity and will reduce the compressive strength [42–44] and could also introduce more weak parts in cement pastes. They are mainly caused by two factors, one of which is the organic carbon and moisture present in diatoms, which is evidenced by the disconnection between different piles of clustered nuclear-shaped HD C-S-H in both OPC and GGBS mixes [45–47]. The other is the micron gap formed by the diatoms themselves, which can be obtained by observing the microstructure visually. These agree well with the previously reported results on compressive strength which showed that diatoms in seawater caused a reduction in the compressive strength of cement paste in both OPC and GGBS mixes, especially after 1 day. In GGBS mixes, there is more ettringite formed [48–50], which could provide more bridging effects between disconnected C-S-H particles [51,52] and thus could partly compensate for the strength loss caused by organic parts in diatoms and the micron gap formed by the diatoms. This explains why the GGBS mixes had a lower rate of strength loss with increasing diatom concentration than the OPC mixes, as in Figure 2.

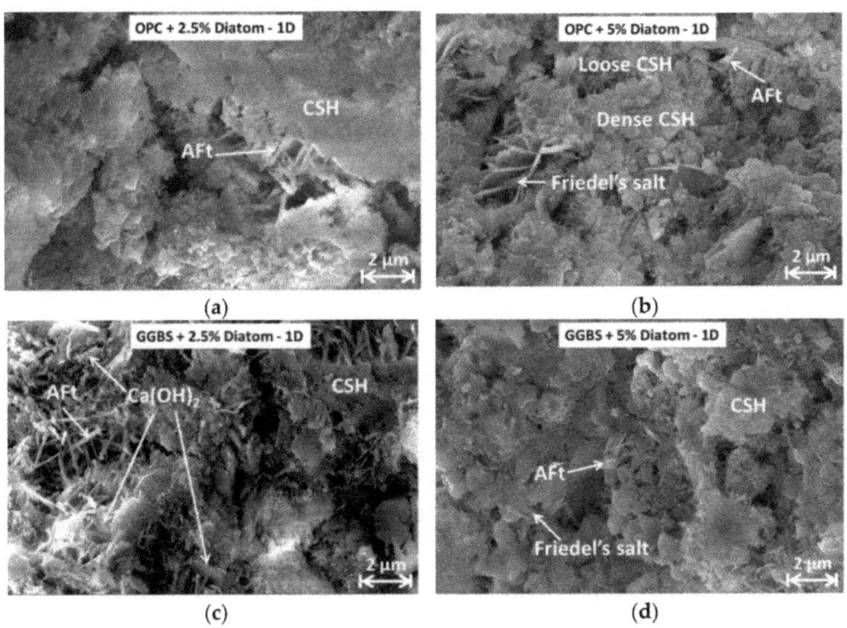

Figure 6. Microstructure observations of different paste samples under SEM at 1D: (**a**) OPC + 2.5% diatom, (**b**) OPC + 5% diatom, (**c**) GGBS + 2.5% diatom, (**d**) GGBS + 5% diatom.

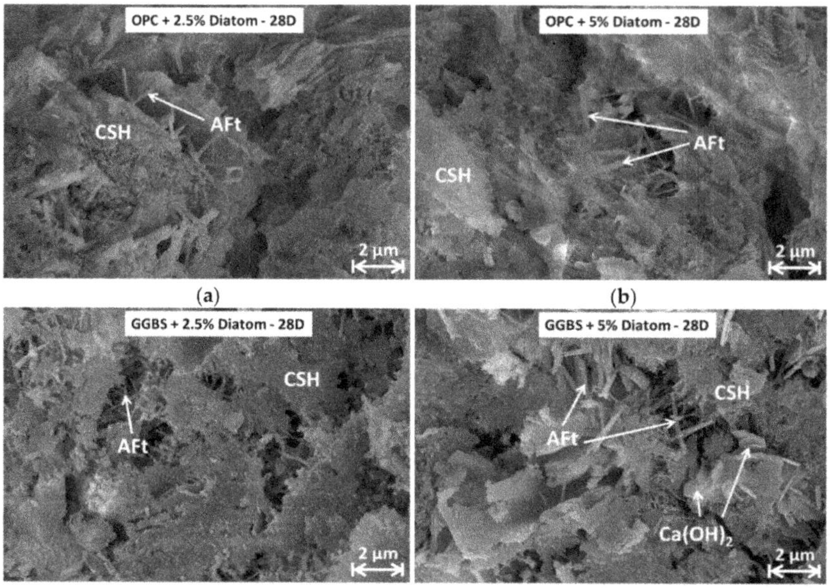

(**a**) (**b**) (**c**) (**d**)

Figure 7. Microstructure observations of different paste samples under SEM at 28D: (**a**) OPC + 2.5% diatom, (**b**) OPC + 5% diatom, (**c**) GGBS + 2.5% diatom, (**d**) GGBS + 5% diatom.

4. Conclusions

The effect of diatom concentrations (0%, 2.5% and 5% by weight of suspension of seawater and diatoms) on the compressive strength, hydration products and microstructure of OPC and GGBS cement pastes was investigated, and the following conclusions can be drawn:

Diatoms in seawater can accelerate the hydration of cement and produce more $Ca(OH)_2$ and C-S-H. The microstructure revealed by SEM indicates that the amorphous nano-SiO_2 and nano-$CaCO_3$ in diatoms could provide the nucleus for C-S-H, and the amorphous nano-SiO_2 also contributes to the pozzolanic reaction in OPC and blended cement paste and thus promotes the hydration of OPC and blended cement.

Although diatoms can accelerate cement hydration, diatoms in seawater are not beneficial for strength development of cement paste, especially after 1 day. This is caused by the organic parts in diatoms and the micron gap formed by the diatoms, which provide weak bonds between C-S-H particles accumulated around the nano-SiO_2, as evidenced by the SEM results.

The reduction in compressive strength of cement paste with 5% diatoms can be up to 50.1% in OPC mixes, and this reduction is much lower in GGBS mixes, with only up to 8.6%. This difference was thought to be due to the fact that more ettringites formed in GGBS mixes than in OPC mixes, and the ettringites in the early ages provide bridging effects between separated C-S-H particles, thus partly compensating for the detrimental effects of the weak bonds between C-S-H particles in GGBS mixes.

Future research is suggested to investigate the long-term performance of concrete mixed with seawater and diatoms, especially on the GGBS mixes. Further, the allowable maximum concentration of diatoms in seawater should be investigated and quantified. The effect of low-concentrated diatoms in seawater on concrete performance needs to be carried out.

Author Contributions: Conceptualization, J.W., W.F. and Y.Z. (Yi Zhu); methodology, J.W., H.H., Y.W. and W.F.; formal analysis, J.W., L.X., J.G., Y.J., H.H., P.H., Y.Z. and Z.Y. (Yi Zhang); investigation, J.W., L.X., J.G., Y.J., H.H. and Z.Y.; writing—original draft preparation, J.W., L.X. and H.H.; writing—review and editing, J.W., Y.W., W.F., P.H., Y.Z. (Yi Zhang) and Y.Z. (Yi Zhu); supervision, J.W. and Y.W.; project administration, J.W.; funding acquisition, J.W. and Y.W. All authors have read and agreed to the published version of the manuscript.

Funding: The author appreciates the financial support from the National Natural Science Foundation of China (52008232). W.F. was funded by the Hundred Talents Programme of Zhejiang University and also supported by Science Foundation of Donghai Laboratory (Grant No. DH-2022KF0203).

Institutional Review Board Statement: Not applicable.

Informed Consent Statement: Not applicable.

Data Availability Statement: The data presented in this study are available on request from the corresponding author.

Conflicts of Interest: The authors declare no conflict of interest.

References

1. Wang, J.J.J.; Liu, E.G.; Li, L. Multiscale investigations on hydration mechanisms in seawater OPC paste. *Constr. Build. Mater.* **2018**, *191*, 891–903. [CrossRef]
2. Wang, J.J.; Xie, J.H.; Wang, Y.L.; Liu, Y.L.; Ding, Y.H. Rheological properties, compressive strength, hydration products and microstructure of seawater-mixed cement pastes. *Cem. Concr. Compos.* **2020**, *114*, 103770. [CrossRef]
3. Xiao, J.; Qiang, C.; Nanni, A.; Zhang, K. Use of sea-sand and seawater in concrete construction: Current status and future opportunities. *Constr. Build. Mater.* **2017**, *155*, 1101–1111. [CrossRef]
4. Bazli, M.; Li, Y.L.; Zhao, X.L.; Raman, R.K.S.; Bai, Y.; Al-Saadi, S.; Haque, A. Durability of seawater and sea sand concrete filled filament wound FRP tubes under seawater environments. *Compos. Part B-Eng.* **2020**, *202*, 108409. [CrossRef]
5. Dhondy, T.; Remennikov, A.; Sheikh, M.N. Properties and Application of Sea Sand in Sea Sand-Seawater Concrete. *J. Mater. Civ. Eng.* **2020**, *32*, 04020392. [CrossRef]
6. Yang, J.L.; Lu, S.W.; Wang, J.Z.; Wang, Z. Behavior of CFRP partially wrapped square seawater sea-sand concrete columns under axial compression. *Eng. Struct.* **2020**, *222*, 111119. [CrossRef]
7. Dasar, A.; Patah, D.; Hamada, H.; Sagawa, Y.; Yamamoto, D. Applicability of seawater as a mixing and curing agent in 4-year-old concrete. *Constr. Build. Mater.* **2020**, *259*, 119692. [CrossRef]
8. Mangi, S.A.; Makhija, A.; Raza, M.S.; Khahro, S.H.; Jhatial, A.A. A Comprehensive Review on Effects of Seawater on Engineering Properties of Concrete. *Silicon* **2021**, *13*, 4519–4526. [CrossRef]
9. Zhang, Y.R.; Wei, Y.; Bai, J.W.; Wu, G.; Dong, Z.Q. A novel seawater and sea sand concrete filled FRP-carbon steel composite tube column: Concept and behaviour. *Compos. Struct.* **2020**, *246*, 112421. [CrossRef]
10. Li, Y.L.; Zhao, X.L.; Raman, R.K.S. Durability of seawater and sea sand concrete and seawater and sea sand concrete-filled fibre-reinforced polymer/stainless steel tubular stub columns. *Adv. Struct. Eng.* **2020**, *24*, 1074–1089. [CrossRef]
11. Chen, G.M.; Liu, P.C.; Jiang, T.; He, Z.B.; Wang, X.M.; Lam, L.; Chen, J.F. Effects of natural seawater and sea sand on the compressive behaviour of unconfined and carbon fibre-reinforced polymer-confined concrete. *Adv. Struct. Eng.* **2020**, *23*, 3102–3116. [CrossRef]
12. Mbadike, E.M.; Elinwa, A.U. Effect of salt water in the production of concrete. *Niger. J. Technol.* **2011**, *30*, 105–110. Available online: https://nijotech.com/index.php/nijotech/article/view/31 (accessed on 1 October 2022).
13. Younis, A.; Ebead, U.; Suraneni, P.; Nanni, A. Fresh and hardened properties of seawater-mixed concrete. *Constr. Build. Mater.* **2018**, *190*, 276–286. [CrossRef]
14. Montanari, L.; Suraneni, P.; Tsui-Chang, M.; Khatibmasjedi, M.; Ebead, U.; Weiss, J.; Nanni, A. Hydration, Pore Solution, and Porosity of Cementitious Pastes Made with Seawater. *J. Mater. Civ. Eng.* **2019**, *31*, 04019154. [CrossRef]
15. De Martino, A.; Bartual, A.; Willis, A.; Meichenin, A.; Villazan, B.; Maheswari, U.; Bowler, C. Physiological and Molecular Evidence that Environmental Changes Elicit Morphological Interconversion in the Model Diatom *Phaeodactylum tricornutum*. *Protist* **2011**, *162*, 462–481. [CrossRef] [PubMed]
16. Tesson, B.; Gaillard, C.; Martin-Jezequel, V. Insights into the polymorphism of the diatom *Phaeodactylum tricornutum* Bohlin. *Bot. Mar.* **2009**, *52*, 104–116. [CrossRef]
17. Moreno, J.J.G.; Pan, K.; Wang, Y.; Li, W.J. Computational Study of APTES Surface Functionalization of Diatom-like Amorphous SiO_2 Surfaces for Heavy Metal Adsorption. *Langmuir* **2020**, *36*, 5680–5689. [CrossRef]
18. Yilmaz, E.U.; Yilmaz, M. Diatom-Derived SiO_2/N-Doped C Nanocomposite for Electrochemical Capacitors. *ECS J. Solid State Sci. Technol.* **2020**, *9*, 061012. [CrossRef]
19. Zhang, D.Y.; Pan, J.F.; Cai, J.; Wang, Y.; Jiang, Y.G.; Jiang, X.G. Hydrofluoric acid-assisted bonding of diatoms with SiO_2-based substrates for microsystem application. *J. Micromech. Microeng.* **2012**, *22*, 035021. [CrossRef]

20. Dudley, S.; Kalem, T.; Akinc, M. Conversion of SiO_2 diatom frustules to $BaTiO_3$ and $SrTiO_3$. *J. Am. Ceram. Soc.* **2006**, *89*, 2434–2439. [CrossRef]
21. Wang, Y.L.; Lu, H.J.; Wang, J.J.; He, H. Effects of Highly Crystalized Nano C-S-H Particles on Performances of Portland Cement Paste and Its Mechanism. *Crystals* **2020**, *10*, 816. [CrossRef]
22. Long, G.; Li, Y.; Ma, C.; Xie, Y.; Shi, Y. Hydration kinetics of cement incorporating different nanoparticles at elevated temperatures. *Thermochim. Acta* **2018**, *664*, 108–117. [CrossRef]
23. Nuaklong, P.; Sata, V.; Wongsa, A.; Srinavin, K.; Chindaprasirt, P. Recycled aggregate high calcium fly ash geopolymer concrete with inclusion of OPC and nano-SiO_2. *Constr. Build. Mater.* **2018**, *174*, 244–252. [CrossRef]
24. Shaikh, F.U.; Supit, S.W.; Salim, B. Microstructure and Nanoscaled Characterization of HVFA Cement Paste Containing Nano-SiO_2 and Nano-$CaCO_3$. *J. Mater. Civ. Eng.* **2017**, *29*, 04017063. [CrossRef]
25. Mohseni, E.; Miyandehi, B.M.; Yang, J.; Yazdi, M.A. Single and combined effects of nano-SiO_2, nano-Al_2O_3 and nano-TiO_2 on the mechanical, rheological and durability properties of self-compacting mortar containing fly ash. *Constr. Build. Mater.* **2015**, *84*, 331–340. [CrossRef]
26. Xu, L.; Wang, J.; Li, K.; Lin, S.; Li, M.; Hao, T.; Ling, Z.; Xiang, D.; Wang, T. A systematic review of factors affecting properties of thermal-activated recycled cement, Resources. *Conserv. Recycl.* **2022**, *185*, 106432. [CrossRef]
27. Pourjavadi, A.; Fakoorpoor, S.M.; Khaloo, A.; Hosseini, P. Improving the performance of cement-based composites containing superabsorbent polymers by utilization of nano-SiO_2 particles. *Mater. Des.* **2012**, *42*, 94–101. [CrossRef]
28. Li, J.; Zhang, W.; Li, C.; Monteiro, P.J.M. Green concrete containing diatomaceous earth and limestone: Workability, mechanical properties, and life-cycle assessment. *J. Clean. Prod.* **2019**, *223*, 662–679. [CrossRef]
29. Aydin, A.C.; Gül, R. Influence of volcanic originated natural materials as additives on the setting time and some mechanical properties of concrete. *Constr. Build. Mater.* **2007**, *21*, 1277–1281. [CrossRef]
30. Degirmenci, N.; Yilmaz, A. Use of diatomite as partial replacement for Portland cement in cement mortars. *Constr. Build. Mater.* **2009**, *23*, 284–288. [CrossRef]
31. Wang, J.; Xu, L.; Li, M.; He, H.; Wang, Y.; Xiang, D.; Lin, S.; Zhong, Y.; Zhao, H. Effect of pre-carbonation on the properties of cement paste subjected to high temperatures. *J. Build. Eng.* **2022**, *51*, 104337. [CrossRef]
32. Wang, J.J.; Xie, J.H.; Wang, C.H.; Zhao, J.B.; Liu, F.; Fang, C. Study on the optimum initial curing condition for fly ash and GGBS based geopolymer recycled aggregate concrete. *Constr. Build. Mater.* **2020**, *247*, 118540. [CrossRef]
33. Xie, J.H.; Liu, J.F.; Liu, F.; Wang, J.J.; Huang, P.Y. Investigation of a new lightweight green concrete containing sludge ceramsite and recycled fine aggregates. *J. Clean. Prod.* **2019**, *235*, 1240–1254. [CrossRef]
34. Wang, J.; Basheer, P.A.M.; Nanukuttan, S.V.; Long, A.E.; Bai, Y. Influence of service loading and the resulting micro-cracks on chloride resistance of concrete. *Constr. Build. Mater.* **2016**, *108*, 56–66. [CrossRef]
35. Alhozaimy, A.; Al-Negheimish, A.; Alawad, O.A.; Jaafar, M.S.; Noorzaei, J. Binary and ternary effects of ground dune sand and blast furnace slag on the compressive strength of mortar. *Cem. Concr. Compos.* **2012**, *34*, 734–738. [CrossRef]
36. He, Z.; Zhu, X.; Wang, J.; Mu, M.; Wang, Y. Comparison of CO_2 emissions from OPC and recycled cement production. *Constr. Build. Mater.* **2019**, *211*, 965–973. [CrossRef]
37. Wang, J.; Liu, E.; Li, L. Characterization on the recycling of waste seashells with Portland cement towards sustainable cementitious materials. *J. Clean. Prod.* **2019**, *220*, 235–252. [CrossRef]
38. Wang, J.J.; Liu, E.G. Upcycling waste seashells with cement: Rheology and early-age properties of Portland cement paste. *Resour. Conserv. Recycl.* **2020**, *155*, 104680. [CrossRef]
39. Wang, Y.; He, F.; Wang, J.; Hu, Q. Comparison of Effects of Sodium Bicarbonate and Sodium Carbonate on the Hydration and Properties of Portland Cement Paste. *Materials* **2019**, *12*, 1033. [CrossRef]
40. Constantinides, G.; Ulm, F.J. The effect of two types of C-S-H on the elasticity of cement-based materials: Results from nanoindentation and micromechanical modeling. *Cem. Concr. Res.* **2004**, *34*, 67–80. [CrossRef]
41. Constantinides, G.; Ulm, F.J.; Van Vliet, K. On the use of nanoindentation for cementitious materials. *Mater. Struct.* **2003**, *36*, 191–196. [CrossRef]
42. Liu, J.; Wu, K.W.; Wang, Y.F.; Yang, Y.Q. Effects of fly Ash/diatomite admixture with variable particle sizes on the mechanical properties and porosity of concrete. *J. Wuhan Univ. Technol. -Mater. Sci. Ed.* **2017**, *32*, 1072–1079. [CrossRef]
43. Yilmaz, B. A study on the effects of diatomite blend in natural pozzolan-blended cements. *Adv. Cem. Res.* **2008**, *20*, 13–21. [CrossRef]
44. Senff, L.; Hotza, D.; Labrincha, J.A. Effect of diatomite addition on fresh and hardened properties of mortars investigated through mixture experiments. *Adv. Appl. Ceram.* **2011**, *110*, 142–150. [CrossRef]
45. Xie, J.; Zhao, J.; Wang, J.; Fang, C.; Yuan, B.; Wu, Y. Impact behaviour of fly ash and slag-based geopolymeric concrete: The effects of recycled aggregate content, water-binder ratio and curing age. *Constr. Build. Mater.* **2022**, *331*, 127359. [CrossRef]
46. Hang, H.; Wang, Y.; Wang, J. Effects of aggregate micro fines (AMF), aluminum sulfate and polypropylene fiber (PPF) on properties of machine-made sand concrete. *Appl. Sci.* **2019**, *9*, 2250.
47. Wang, J.; Xie, J.; He, J.; Sun, M.; Zhao, J. Combined use of silica fume and steel fibre to improve fracture properties of recycled aggregate concrete exposed to elevated temperature. *J. Mater. Cycles Waste Manag.* **2020**, *22*, 862–877. [CrossRef]
48. Wang, J. Steady-state chloride diffusion coefficient and chloride migration coefficient of cracks in concrete. *J. Mater. Civ. Eng.* **2017**, *29*, 0417117. [CrossRef]

49. Wang, J.; Liu, E. The relationship between steady-state chloride diffusion and migration coefficients in cementitious materials. *Mag. Concr. Res.* **2020**, *72*, 1016–1026. [CrossRef]
50. Xie, J.; Zhao, J.; Wang, J.; Huang, P.; Liu, J. Investigation of the high-temperature resistance of sludge ceramsite concrete with recycled fine aggregates and GGBS and its application in hollow blocks. *J. Build. Eng.* **2021**, *34*, 101954. [CrossRef]
51. Li, J.Q.; Jin, Q.X.; Zhang, W.X.; Li, C.; Monteiro, P.J.M. Microstructure and durability performance of sustainable cementitious composites containing high-volume regenerative biosilica. *Resour. Conserv. Recycl.* **2022**, *178*, 106038. [CrossRef]
52. Liu, R.Q.; Yang, Y.Q.; Zhao, X.K.; Pang, B. Quantitative phase analysis and microstructural characterization of Portland cement blends with diatomite waste using the Rietveld method. *J. Mater. Sci.* **2021**, *56*, 1242–1254. [CrossRef]

coatings MDPI

Article

Effectiveness of Bio-Waste-Derived Carbon Doping on De-Icing Performance of an Electrically Resistant Concrete

Baglan Bakbolat [1,2], Chingis Daulbayev [3,4], Fail Sultanov [1,2,3,*], Azamat Taurbekov [1,2], Aidos Tolynbekov [2], Mukhtar Yeleuov [2,5], Alina V. Korobeinyk [6] and Zulkhair Mansurov [1,2]

1 Faculty of Chemistry and Chemical Technology, Al-Farabi Kazakh National University, Almaty 050040, Kazakhstan
2 Institute of Combustion Problems, Almaty 050000, Kazakhstan
3 National Laboratory Astana, Nazarbayev University, Nur-Sultan 010000, Kazakhstan
4 Institute of Nuclear Physics, Almaty 050032, Kazakhstan
5 Department of Materials Science, Nanotechnology and Engineering Physics, Satbayev University, Almaty 050000, Kazakhstan
6 Aarhus Institute of Advanced Studies, Aarhus University, 8000 Aarhus C, Denmark
* Correspondence: fail_23@bk.ru; Tel.: +7-7014222293

Abstract: This paper proposes a modified carbon-based concrete filler composition, which can potentially be used as a self-de-icing pavement. Carbon fibers (CNFs), graphene-like porous carbon (GLC), and a CNF/GLC composite were developed to reinforce concrete with the aim to improve its electrical conductivity and mechanical properties. The effect of the CNF and GLC loadings on the electrical conductivity of the filled concrete was evaluated in a climatic chamber at temperatures simulating water-freezing conditions on a concrete road. The results show that even a negligible loading (0.2 wt.%) of concrete with CNF/GLC results in a dramatic decrease in its resistance when compared to the same loadings for CNF and GLC added separately. Depending on the number of fillers, the temperature of the modified concrete samples reached up to +19.8 °C at low voltage (10 V) at −10 °C, demonstrating the perspective of their heat output for anti-icing applications. Additionally, this study shows that adding 2.0 wt.% of the CNF/GLC composite to the concrete improves its compressive strength by 33.93% compared to the unmodified concrete.

Keywords: carbon fiber; bio-waste graphene-like porous carbon; modified concrete; de-icing; electrical resistance heating

1. Introduction

The icing of highways, bridges, airport runways, and other transport infrastructures poses a serious danger to people's lives. According to recent studies, snow and ice formation are responsible for more than 15% of all road traffic accidents, which leads to huge human and economic losses [1–4]. The traditional de-icing methods are predominantly based on mechanical cleaning and the utilization of various chemicals such as sodium, calcium, and magnesium chlorides. Though these methods have found wide applications due to their low cost and simplicity, their long-term negative impact on the condition of road surfaces and their contribution to environmental pollution are undeniable facts. The use of salts leads to the formation of cracks in concrete and causes the corrosion of the steel reinforcement inside the concrete. Meanwhile, the annual use of salts leads to their accumulation in the soil, which negatively affects the environment. Studies [5–11] indicate that salts used for road de-icing are causing water deterioration and the inhibition of tree growth along roadsides.

The alternative method to reduce the icing of road surfaces is via locally heating the concrete by passing an electric current through said surfaces. Since concrete is a dielectric, various additives with high electrical conductivity are commonly used. Steel shavings and wool are widely used as such additives, and their content of up to 10 vol.% enables the achievement of the required values of conductivity and the mechanical hardness of the composite concrete [12,13]. Recent developments in additives undertaken to enhance their electrical conductivity indicate that their size and distribution within the concrete matrix are of decisive importance compared to the total mass loading [14]. Authors [12,15] have demonstrated the importance of such parameters as the micro- and nano-sizes of the filler particles and the control of the filler/cement, water/cement, and sand/cement ratios in the achievement of the desired electrical resistance of the resulting concrete.

The benefits of carbon materials such as graphene [15,16], CNFs [17], and и-activated carbon [18,19] towards increasing concrete's electrical conductivity have also been reported. Compared to steel wool and shavings, carbon materials do not require a large loading, and the resulting carbon material-modified concrete does not require high electric current, which reduces the potential risk to pedestrians and vehicles. However, the addition of graphene or carbon nanotubes, even at less than 1 wt.%, significantly increases the cost of the modified concrete. Previously, we successfully demonstrated that a graphene-like material obtained via the carbonization and thermo-chemical activation of biological wastes (rice husks and apricot shells) is characterized as quite comparable to commercial graphene's optical, mechanical, and electrical properties [20–22]. The utilization of an inexpensive approach in the synthesis of targeted valuable carbon-based products from bio-waste precursors offers great advantages since it reduces the cost of the final nanostructured material significantly. The addition of graphene initially increases the electrical conductivity of concrete, which can be further improved by using CNF-based additives. Moreover, CNFs not only provide a high electrical conductivity, but also serve as a reinforcing agent in the structure of concrete [23–25].

Considering the above, the purpose of this study is to develop an additive to improve the effectiveness of concrete de-icing to suppress road surface icing and thereby solve the problems faced in road maintenance. To achieve the research objectives, the concrete has been modified with carbon nanofibers and a graphene-like porous carbon obtained from biowaste. The utilization of bio-waste materials in the carbonaceous filler's synthesis is aimed at reducing the cost of electrically conductive concrete [26]. Furthermore, a simple and inexpensive electrospinning method was used for the carbon nanofibers' production [27]. The influence of additives on the mechanical properties on the resulting concrete was evaluated via measurements of the specific resistance of concrete blocks. According to the obtained results, the optimal composition of the concrete mixture modified with carbon filler (1 wt.%) not only improves the values of conductivity but also enhances the mechanical characteristics of the concrete structure and its hydrophobicity. As a result of the gained hydrophobicity, the concrete will adsorb less water (and fewer dissolved substances) and, consequently, will be better protected during exploitation.

2. Materials and Methods

2.1. Synthesis of Graphene-like Porous Carbon from Biological Waste

Graphene-like porous carbons (GLC) were obtained from the bio-waste with reference to methods reported elsewhere [20,27] with minor changes. Briefly, pre-cleaned walnut shells (WSs) were ground to an average size of 2–6 mm using an impact mill; then, they were thoroughly washed with hot distilled water and passed over a 2 mm sieve to remove smaller particles and dust and dried at 120 °C (with a heating rate of 10 °C min^{-1}) for 10 h in a bench oven. Afterwards, crushed WSs were carbonized at 480 °C, with a heating rate of 5 °C min^{-1}, for 110 min, and under Ar atmosphere with a gas flow rate of 160 cm^3/min. Further, the carbonized WSs were thermochemically activated by mixing them with KOH (Sigma Aldrich, Saint Louis, MO, USA, ≥90%, flakes) at 1:4 mass ratio, and then heating them for 5 h at 360 °C to execute the complete impregnation of carbonized WSs with KOH. The carbonaceous product impregnated with KOH was then thermo-chemically activated at 850 °C over 90 min (heating rate 7 °C min^{-1}; Ar flow rate of 220 cm^3/min). Obtained carbonaceous product was washed with hot deionized water until neutral pH was reached for complete removal of reaction by-products. Finally, as-received GLC was dried in a bench oven at a temperature of 120 °C for 10 h and then in a vacuum oven at 150 °C and pressure of 10^{-2}–10^{-3} Torr for 2 h.

2.2. Electrospinning of CNF/GLC Composite

To produce 1D composite of GLC and CNFs, the electrospinning method followed by two-step carbonization was applied. A total of 0.5 g of polyacrylonitrile (PAN) was dissolved in 10 mL of ethanol and 0.05 g of GLC was added to the mixture under constant stirring (120 rpm) for 30 min, ensuring the complete homogenization of GLC within the polymer solution. Resulting solution was then loaded into a 2.5 mL medical syringe and fibers were formed via an electrospinning setup (voltage—30 kV and distance between needle and collector—15 cm), the scheme of which is given Figure 1a. Collected electrospun GLC/PAN mats were thermally stabilized at 185 °C for 15 min, calcined at 500 °C for 30 min in Ar atmosphere, and designated as CNF/GLC composites. GNFs represent carbon fibers obtained from PAN fibers without carbon filler.

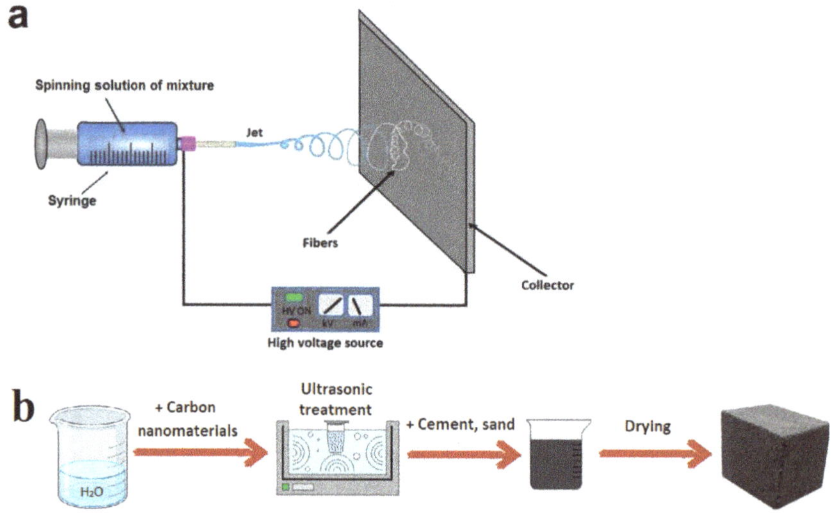

Figure 1. (**a**) Scheme of electrospinning setup with a needle and flat horizontal metal collector, and (**b**) preparation of concrete samples modified with the carbonaceous additive.

2.3. Preparation of the Concrete Filled with CNF/GLC Composite

Prior to the sample preparation of filled concrete, an appropriate amount of filler (CNFs, GLC, or CNF/GLC composite) was weighed and dispersed ultrasonically in DI water (15 mL) at 32 kHz for 30 min (Figure 1b). Portland Cement (M300, Type II, ASTM C 150) was used in the concrete mix; it has a specific gravity of 2.9, requires 34% of its volume to be water to obtain cement paste of standard consistency, and has an initial setting time of about 45 min. Good quality river sand with specific gravity of 2.7 (BSI BS 882:1992) was used as a fine additive. The optimal ratio for concrete samples' preparation was experimentally determined to be 1:1.5:0.9 (cement:sand:aqueous solution of filler); fillers were added as aqueous dispersion ranging from 0.2 to 2.4 wt.% (with 0.2 wt.% step) for final composition. It was found that blank concrete mix required less water to obtain cement paste of standard consistency, and concrete blank mix design is as follows: 1:1.5:0.8 (cement:sand:water). This is probably due to the specific mass–volume ratio of carbonaceous fillers. Test specimens were casted as 20 mm × 20 mm × 20 mm cubes per mix proportion with and without fillers to determine their electrical resistance and compressive strength. For each fiber proportion, 3 cubes were cast for each trial. In total, 111 cube-shaped samples were cast for the complete investigation. The specimens were de-molded after 24 h and then cured for 28 days in desiccators, at ambient temperature and partially submerged conditions.

2.4. Characterization

The surface morphology of the obtained carbonaceous samples and their composites was investigated using a Quanta 3D 200i scanning electron microscope (SEM, FEI, USA) with an accelerating voltage of 15 kV. The structure of the GLC powder was investigated by transmission electron microscopy (TEM, JEM-2100, Japan) with high-voltage and current stability. Structural analysis was performed using the X-ray powder diffraction data obtained via DRON-8 diffractometer (Cu Kα-radiation). Data were collected with scanning rate of 0.02° in a 2θ range from 0 to 70°. A Raman characterization was performed on a Raman spectrometer NTEGRA Spectra (λ = 473 nm; the area signals had a diameter of 80 nm). Compressive strength characteristics of filled and unfilled concrete samples were tested with Shimadzu AGX 100 kN electromechanical-testing machine at a crosshead speed of 0.5 mm/min. Surface-wetting angles of the concrete samples were measured by a goniometer; in all wettability tests, 2 μL of distilled water was placed on each sample's surface using a handy step pipette.

2.5. Electrical Conductivity and Heating Performance Tests

To determine the electrical resistance of cubic concrete samples, a two-probe method was used. Figure 2 presents the scheme and a digital image of experimental cell setup for measurement of electrical resistance of concrete cubes using a digital multimeter, Total (Model: TMT47503). The cell was made from a Teflon (dielectric) and a constant voltage of 10 V was applied to the copper plate electrodes at both sides of the concrete blocks via DC power supply (UTP3315TFL-II). The concrete blocks' heating rate at negative temperatures was studied in climatic chamber at conditions simulating ice formation on a road using thermal image-capturing technique with a Bosch GTC 400 C thermal camera. The additional specifications of the used equipment are presented in Table 1.

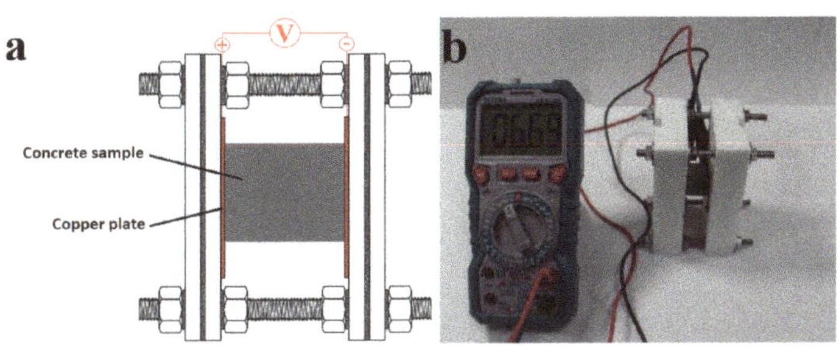

Figure 2. (**a**) The scheme of the cell setup for measurement of the electrical resistance of cement specimen and (**b**) optical image of the cell setup with connected digital multimeter.

Table 1. Specifications of measurement equipment used during the test.

Equipment	Manufacturer/Model	Specifications
Digital multimeter	Total (TMT47503)	Non-contact voltage detection; DC Voltage: 600 mV/6 V/60 V/600 V/1000 V ± (0.5% + 3); Resistance: 600 Ω/6 kΩ/60 kΩ/600 kΩ/6 MΩ ± (0.8% + 3).
Thermal camera	Bosch GTC 400 C	Measurement range: 10 °C to + 400 °C; Measurement accuracy of IR: ±3.0 °C; Resolution: 0.1 °C; Focus distance—minimum: 0.3 m.

3. Results and Discussion

3.1. CNF, GLC, and CNF/GLC Composite Characterization

SEM images of the WS-derived carbonaceous products before and after chemical activation are shown in Figure 3. As can be seen, the surface of the WS after carbonization is represented as a rough, dense structure with defects and small indentations (Figure 3a–c). Thermo-chemical activation transformed the WS into a flaky, spongy, and porous GLC structure (Figure 3d–f). This significant surface morphology alteration is explained by the harsh conditions of the thermal transformation/activation when the evacuation of low-volatility compounds results in the formation of new structural arrangements completely different from the initial ones. The TEM analysis of the WS-derived structures (Figure 3g,h) revealed the presence of 2D graphene-like nanosheets of a randomly layered structure in the prepared GLC. Wrinkles and overlapping layers (darker area in the TEM images) are characteristic of biomass-derived graphene and contribute to the porous structure of GLC [28], while the light areas in the TEM images indicate the presence of sparsely layered graphene in the structure [29].

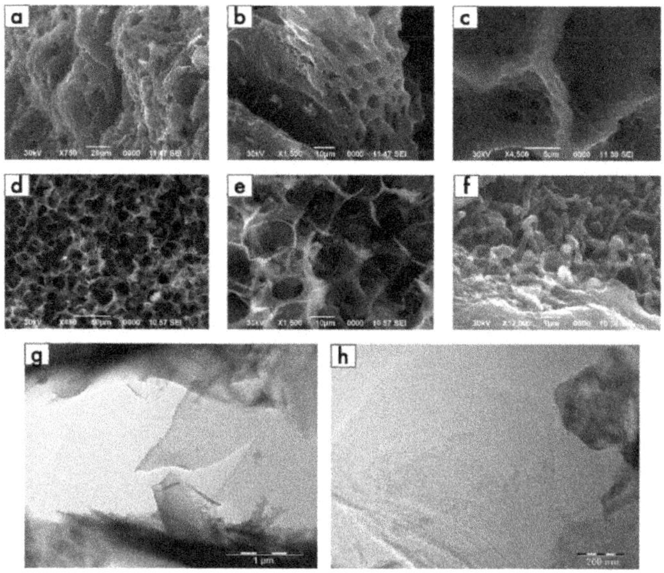

Figure 3. Electron microphotographs of WS-derived carbons: (**a–c**) SEM images before thermo-chemical activation; (**d–f**) SEM and (**g,h**) TEM images after thermo-chemical activation.

The Raman spectrum of the GLC (Figure 4a) contains both the G band (1582 cm^{-1}) and the D-band (1357 cm^{-1}). The D-band is related to the edge and structural defects of disordered carbon, while the G-band corresponds to sp^2-hybridized carbon atoms, confirming the presence of graphite in the sample. The 2D-band (2717 cm^{-1}) can be used for the calculation of the number of graphene layers in the structure of the GLC [30,31]. According to the ratio of I_g to I_{2D}, the obtained GLC has a sparsely layered (5–7 layers) graphene structure. The graphene-like structure of the GLC was also confirmed through the XRD analysis. The XRD pattern of the GLC sample (Figure 4b) contains a diffraction peak at 25.87 2θ° (002), which corresponds to the formation of a graphene-like structure [32]. A broad peak of low intensity at 43.25 2θ° (100) indicates the absence of a graphitized phase in the sample. The calculated interplanar spacing (d) for the obtained GLC samples is 0.34405 nm, which is close to the theoretical spacing of graphene nanosheets and is higher than that for graphite.

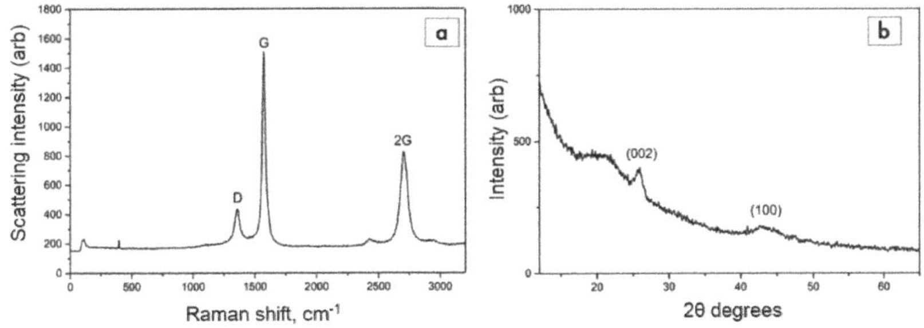

Figure 4. (**a**) Raman spectra and (**b**) XRD pattern of GLC.

To fabricate the CNFs and CNF/CLC composite, the electrospinning method, followed by the thermal stabilization and calcination of the 1D samples, was used. The structural and morphological characteristics of the CNF and CNF/GLC fibers were investigated by SEM and TEM. The CNF sample in the SEM image (Figure 5a) appears as network of fibers with an average diameter up to 100 nm and a random orientation. In the case of the CNF/GLC composite (Figure 5b), some agglomeration and formation of additional layers along the CNF surface can be observed. These agglomerates are probably caused by the presence of GLC in solution during the electrospinning process. A further investigation of the interaction nature between the CNF and GLC within the composite was performed using TEM. According to the TEM images of the CNF/GLC composite (Figure 5c,d), the nanoplatelets of GLC are homogeneously distributed within the CNF structure. Thus, the successful formation of a homogeneous CNF/GLC composite was confirmed.

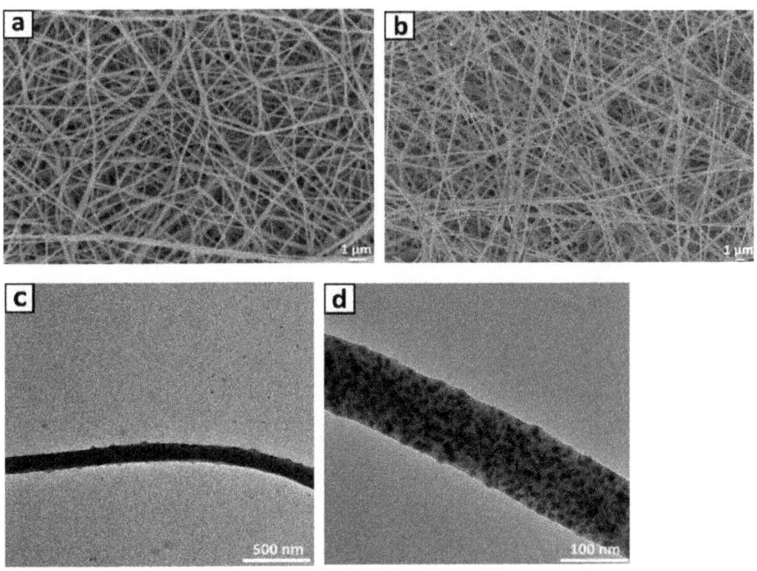

Figure 5. (**a**) SEM images of CNF and (**b**) CNF/GLC composite and (**c**,**d**) TEM images of CNF/GLC composite.

3.2. Investigation of the Concrete Unfilled and Filled with CNFs, GLC, and CNF/GLC

3.2.1. Electrical Resistance Tests

The addition of carbonaceous materials into the concrete mix considerably altered the concrete's electrical resistance. The introduction of 0.2 wt.% of the CNF/GLC composite into the concrete mix resulted in a sharp decrease in resistance (8.4 ± 0.29 MΩ), which is rather low when compared with the same loading of CNFs (14.1 ± 0.2 MΩ) or GLC (13.9 ± 0.14 MΩ) (Table 2). However, increasing the concentration of the CNF/GLC composite above 1.0 wt.% in the concrete mixture did not lead to an increase in the values of electrical resistance, and when compared to the same values of the CNFs and GLC, the differences in the values of electrical resistance become indiscernible.

Table 2. Electrical resistance of concrete after addition of carbon nanomaterials *.

No.	CNF			GLC			CNF/GLC Composite		
	Content, wt.%	Resistance, MΩ	Power, W	Content, wt.%	Resistance, MΩ	Power, W	Content, wt.%	Resistance, MΩ	Power, W
1	0.2	14.1 ± 0.2	7.09	0.2	13.9 ± 0.14	7.19	0.2	8.4 ± 0.29	11.9
2	0.4	6.3 ± 0.87	15.87	0.4	5.5 ± 0.78	18.18	0.4	3.2 ± 0.76	31.25
3	0.6	2.9 ± 0.15	34.48	0.6	4.3 ± 0.22	23.26	0.6	1.2 ± 0.6	83.33
4	0.8	1.7 ± 0.3	58.82	0.8	1.2 ± 0.21	83.33	0.8	0.8 ± 0.065	125
5	1	0.42 ± 0.12	238.1	1	0.36 ± 0.12	277.78	1	0.29 ± 0.045	344.83
6	1.2	0.38 ± 0.046	263.16	1.2	0.45 ± 0.069	222.22	1.2	0.31 ± 0.04	322.58
7	1.4	0.381 ± 0.17	262.47	1.4	0.44 ± 0.17	227.27	1.4	0.3 ± 0.17	333.33
8	1.6	0.34 ± 0.01	294.12	1.6	0.21 ± 0.096	476.19	1.6	0.31 ± 0.1	322.58
9	1.8	0.35 ± 0.021	285.71	1.8	0.26 ± 0.078	384.62	1.8	0.22 ± 0.02	454.55
10	2	0.39 ± 0.019	256.41	2	0.15 ± 0.042	666,67	2	0.1 ± 0.019	1000
11	2.2	0.42 ± 0.084	238.1	2.2	0.14 ± 0.035	714.29	2,2	0.12 ± 0.09	833.33
12	2.4	0.37 ± 0.014	270.27	2.4	0.12 ± 0.017	833.33	2.4	0.11 ± 0.041	909.09
	CNF			GLC			CNF/GLC composite		
	Content, wt.%	Resistance, MΩ	Power, W	Content, wt.%	Resistance, MΩ	Power, W	Content, wt.%	Resistance, MΩ	Power, W
	0.2	14.1 ± 0.2	7.09	0.2	13.9 ± 0.14	7.19	0.2	8.4 ± 0.29	11.9
	0.4	6.3 ± 0.87	15.87	0.4	5.5 ± 0.78	18.18	0.4	3.2 ± 0.76	31.25
	0.6	2.9 ± 0.15	34.48	0.6	4.3 ± 0.22	23.26	0.6	1.2 ± 0.6	83.33
	0.8	1.7 ± 0.3	58.82	0.8	1.2 ± 0.21	83.33	0.8	0.8 ± 0.065	125
	1	0.42 ± 0.12	238.1	1	0.36 ± 0.12	277.78	1	0.29 ± 0.045	344.83
	1.2	0.38 ± 0.046	263.16	1.2	0.45 ± 0.069	222.22	1.2	0.31 ± 0.04	322.58
	1.4	0.381 ± 0.17	262.47	1.4	0.44 ± 0.17	227.27	1.4	0.3 ± 0.17	333.33
	1.6	0.34 ± 0.01	294.12	1.6	0.21 ± 0.096	476.19	1.6	0.31 ± 0.1	322.58
	1.8	0.35 ± 0.021	285.71	1.8	0.26 ± 0.078	384.62	1.8	0.22 ± 0.02	454.55
	2	0.39 ± 0.019	256.41	2	0.15 ± 0.042	666.67	2	0.1 ± 0.019	1000
	2.2	0.42 ± 0.084	238.1	2.2	0.14 ± 0.035	714.29	2.2	0.12 ± 0.09	833.33
	2.4	0.37 ± 0.014	270.27	2.4	0.12 ± 0.017	833.33	2.4	0.11 ± 0.041	909.09

* For the control sample without the addition of carbon nanomaterials, the resistance was 15 ± 0.81 MΩ.

The electrical resistance values of the concrete blocks filled with CNF, GLC, or the CNF/GLC composite are presented as a plot function of resistance vs carbonaceous filler loading (Figure 6a). The data analysis confirms that the addition of the CNF/GLC composite at a low amount (0.6 wt.%) significantly affects the electrical resistance of the concrete compared to the control sample (unfilled concrete), for which the resistance value was recorded as 15 ± 0.81 MΩ. The obtained values of electrical resistance for concrete with CNF and GLC content from 0.1 to 2 wt.% are in full agreement with the results of works devoted to the study of various materials, including graphene and CNFs, for electrically conductive concrete preparation [33–35]. However, the addition of carbon nanomaterials to the concrete mix at an amount higher than 1 wt.% is not economically viable. According to [36], the addition of 0.5 wt.% CNFs reduces the cost of electrically conductive concrete by 43% compared to the same for nano-sized carbon (at 6 wt.% loading), and by 37% compared to that of recycled metal shavings (at 1.5 wt.% loading). As a result, the addition of the CNF/GLC composite has a more pronounced effect on the electrical properties of the concrete sample.

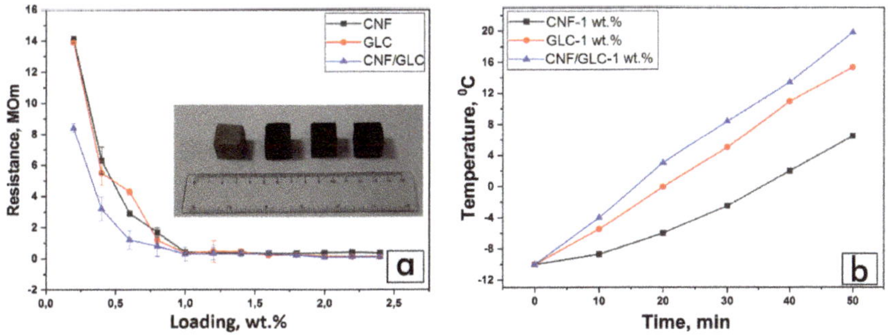

Figure 6. Plots of electrical resistance as function of filler loading (**a**) and concrete specimen temperatures as function of electric current passage time (**b**) for concrete samples with different carbon additives at 1 wt.% filling rate.

Obviously, to determine the optimal mass loading of the carbon-based additives, additional long-term studies focused on electrical conductivity (more than 2 years) are required. The results of these long-term investigations would help to determine the dependence of electric resistance on time and evaluate the effects of CNFs, GLC, and the CNF/GLC composite on the corrosion resistance of electrically conductive concrete. As suggested above, the content of carbon materials in the concrete mixture should not exceed 1 wt.%. As shown, the further increase in the amount of added carbon filler to the concrete did not improve its electrical conductivity; instead, it could negatively affect the cost of electrically conductive concrete. Considering the results obtained in the current investigation, further tests of heat performance for the filled concrete specimens were executed for blocks with a 1 wt.% filling.

3.2.2. Heat Performance Tests

Regarding the electric heating-based thermal de-icing approach, the thermal properties of concrete directly depend on its electrical conductivity. Therefore, to investigate the properties of conductive concrete, it is necessary to measure its electrical conductivity and, based on these results, reveal the optimal parameters for the heating of concrete blocks.

The heating efficiency tests of the concrete blocks with CNFs, GLC, and the CNF/GLC composite (1.0 wt.%) were performed in a climate chamber simulating icing conditions at −10 °C. The increase in the temperature of the concrete blocks was achieved by resistive heating at a voltage of 10 V and was captured by a thermal camera (Figure 7). Each experiment was continued for 50 min with data recorded at 10 min intervals. Data were then plotted as a function of the concrete specimen's temperature vs. the time of the experiment (Figure 6b). It was found that at over 50 min of experiment time in the climate chamber, the concrete blocks filled with 1 wt.% of CNFs, GLC, and the CNF/GLC composite reached 6.5 °C, 15.3 °C, and 19.8 °C (Figure 6b), respectively, from an initial temperature of −10 °C at the zero moment of the experiment (Figures 6b and 7). Meanwhile, the temperature of the concrete sample without the addition of carbon materials remained unchanged.

It is worth noting that the heating occurred evenly in every filled concrete sample under investigation, which indicates the appropriate selection of the mixing parameters of concrete and carbonaceous fillers. According to the infrared image analysis (Figure 7), the temperature was distributed within the block volumes evenly, while the contact points of the concrete block and copper plates do not create a temperature contrast compared to the other areas, confirming a good affinity between the concrete and copper pads. Since the good surface contact between the copper and the concrete block does not interfere with the passage of the electric current, the contact resistances are minimized in the places where local overheating occurs. During the construction of concrete road transport systems, it may be preferable to use thick copper plates with a conductive gel, which will create uniform contact and distribute the pressure evenly.

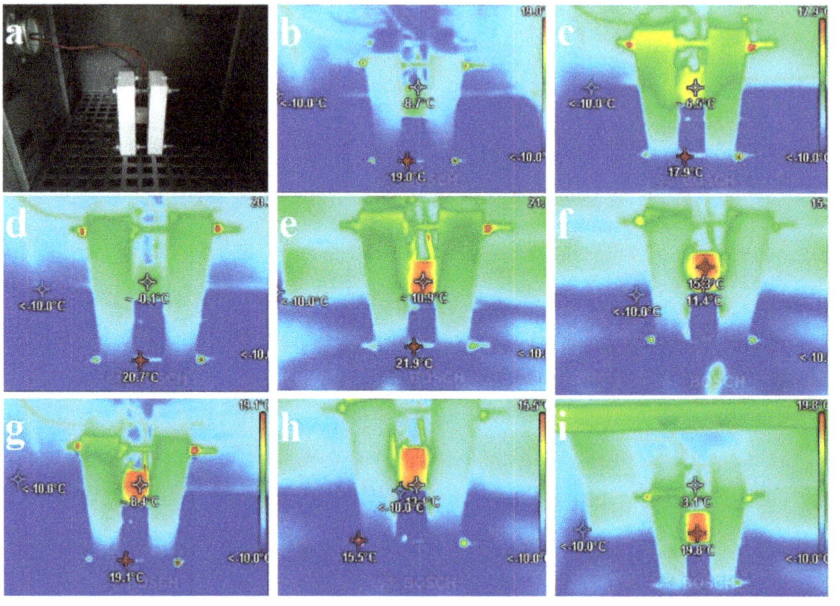

Figure 7. Digital images of the experimental cell setup within the climatic chamber (**a**), and thermal images of concrete blocks filled with CNFs (**b**,**c**), GLC (**d**–**f**), and CNF/GLC composite (**g**–**i**).

3.2.3. Mechanical and Hydrophobic Properties

The influence of the carbon-based additives on the mechanical strength of concrete was investigated by an analysis of the corresponding compressive strengths. It was found that the compressive strength of the concrete specimens increases with an increasing CNF/GLC content. The concrete samples with a content of 1 wt.% and 2 wt.% of the CNF/GLC composite exhibited compression strength values of 23.28 and 33.0 N/mm^2 (Figure 8a), respectively, while this value for the unfilled concrete sample was found to be 21.8 N/mm^2 (Figure 8a). This can be explained by the uniform distribution of the CNF/GLC within the concrete matrix, which results in a strong interaction of the carbon composite with the cement hydration products, leading to the improved bonding of each part of the concrete matrix. Therefore, when the sample is subjected to external forces, the CNF/GLC in the concrete matrix can inhibit and absorb crack propagation energy, resulting in the increased mechanical strength of the concrete. However, the further increase in the carbon-based composite's content in the concrete leads to its possible sticking and agglomeration, which complicates its uniform distribution over the concrete matrix during the preparation procedure. These agglomerates have a negative impact on the possibility of the practical application of the prepared concrete mixture, since numerous mechanically unstable zones are formed in the concrete matrix, leading to a decrease in the concrete's mechanical strength. The wear resistance of concrete is the ability of the concrete surface to resist abrasion. It is known that the use of concrete as a road surface and industrial floors leads to its abrasion, scratching, and slippage due to constant mechanical loads, which result in its surface wear [37,38]. To determine the degree of wear resistance of the concrete, the loss of mass per unit area of the sample and the decrease in the height of the sample subjected to the test were considered. The mechanical tests demonstrated that the concrete sample with 1 and 2 wt.% of the CNF/GLC loadings exhibited abrasion values of 0.93 and 0.98 g/cm^2, while for the carbon-free sample this value was established as 0.87 g/cm^2.

The obtained results confirm that the introduction of carbon composite into a concrete mix leads to a noticeable increase in its wear resistance characteristics.

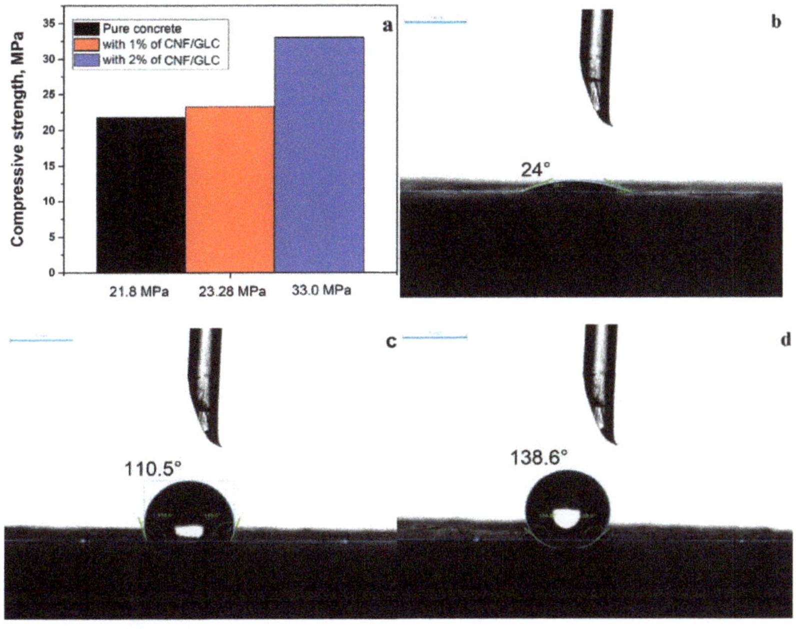

Figure 8. (**a**) Compression strength of CNF/GLC (1 and 2 wt.%) modified concrete blocks compared with pristine concrete block and (**b**) surface-wetting angles of pristine concrete block and (**c**) concrete modified with 1 wt.% and (**d**) 2 wt.% of CNF/GLC.

Hydrophobicity is a vitally important characteristic of coating applications. It has been reported that concrete hydrophobization can be implemented into the manufacture of pavement, as it prevents ion permeability and freeze-thawing leading to the concrete structure's distortion. In addition, the surface-wetting contact angle in concrete (θ) is an important parameter to define the concrete's hydrophobicity or hydrophilicity. It is generally accepted that if $\theta < 90°$, the material's surface is more hydrophilic, and, therefore, water can easily penetrate it; if $\theta > 90°$, the tested surface is more hydrophobic, and water droplets evaporate off of the surface rather than penetrate the capillary. To investigate the influence of the CNF/GLC additive (which has a superior result in the heat performance tests among the filled concrete blocks) on the degree of hydrophobicity of the tested concrete blocks, the surface-wetting angles for the tested specimens were measured and analyzed. Figure 8b–d display the measured results regarding the water contact angles for both the filled and unfilled concrete samples with different CNF/GLC contents. Shown in Figure 8b–d, the wetting angle increased firstly with the CNF/GLC's addition and then increased further with the further increase in the CNF/GLC content. Although the unfilled concrete is intrinsically hydrophilic (Figure 8b, $\theta = 24°$), it reacts with the filler and then forms a structure with hydrophobic properties. The addition of 1 wt.% of the CNF/GLC composite into the concrete matrix resulted in the increase in the block's wetting angle to 110.5° (Figure 8c). The further increase in the carbon composite loading in the concrete mixture (2 wt.%) resulted in an increase in the wetting angle up to 138.6° (Figure 8d), indicating the strong hydrophobic properties of the final material. Thus, the addition of the CNF/GLC composite at a loading of 1 wt.% into the concrete mix results in a significant increase in its surface hydrophobicity. In turn, the added ability of the concrete filled with the CNF/GLC composite to conduct an electric current and heat up, along with the hydrophobicity of its surface, have a significant positive effect on its anti-icing behavior and water penetration resistance.

Table 3 summarizes the results of recent works devoted to the fabrication of self-heating concrete. From the Table, it can be seen that the most commonly used additives in concrete mixes are CNFs, inorganic compounds, steel-based fibers, graphite, etc. The de-icing method is commonly divided into electrothermal and microwave-assisted approaches. The detailed analyses of the heating outputs of the modified concrete types are provided in the Table.

Table 3. Thermal method in concrete de-icing.

Material Composition	Applied De-Icing Approach	Characteristics of the De-Icing Process	Ref.
Carbon fiber conductive bonding layer	Electrothermal	At the constant temperature of −5 °C, the surface temperature of the pavement could reach above 0 °C after the conductive bonding layer was electrified for 100 min under 36 V voltage	[39]
SiC-Fe_3O_4	Microwave	The ice-thawing time of microwave-enhanced functional layer with −15 °C initial temperature was 46 s. It was increased about 1.7 times in contrast to −5 °C.	[40]
Carbon fiber	Microwave	The results show that carbon fiber-modified concrete has the highest heating rate, with a value of 1.680 °C/s, which is 4.46 times higher than that of pristine concrete.	[17]
Steel fiber confined graphite	Electrothermal	The average surface power density of the ramps was between 200 W/m^2 and 300 W/m^2, which is adequate for melting ice in winter environment.	[41]
Carbon fiber heating wires	Electrothermal	A heating experiment on concrete slabs with carbon fiber heating wires was conducted in a refrigerator at −25 °C. It is shown that with an input power of 1134 W/m^2, the temperature on the slab's surface rises above 0 °C after 2.5 h at an approximate rate of 0.17 °C/min.	[42]
Iron black, silicon carbon, and graphite	Microwave	The enhancements of the three absorbing materials ranked from high to low are in the following order: graphite, iron black, and silicon carbon.	[43]
Carbon fibers/graphene-like porous carbon composite (CNF/GLC)	Electrothermal	Modified concrete was heated to +19.8 °C at low voltage (10 V).	This article

4. Conclusions

This research aimed at providing a design approach for de-icing pavement concrete with optimized electrical conductivity and mechanical properties for a critical performance evaluation. The structures of the carbonaceous fillers (CNF and GCL) and their composite (CNF/GCL) as well as a concrete mix were designed, and the efficiency of electrical resistance de-icing was studied. The application value of CNF/GCL in the electrical resistance de-icing of concrete was revealed. From the experimental observations, it has been concluded that:

1. The carbonaceous additive incorporated in the concrete improved its compressive strength, wetting angle, and electrical conductivity compared to the pristine concrete when 0.5 wt.%–1 wt.% was added. The obtained results indicate that the addition of 2.0 wt. % CNF/GLC composite in mortar increases compressive strength by 33.93% compared to unfilled mortar.

2. The addition of 1.0 wt.% carbonaceous additive to the mixture results in an optimum conductive blend for de-icing performance. However, the effect of the CNF/GLC composite is more substantial that the CNF and GLC in terms of performance improvement. The temperature of the modified concrete increases to +19.8 °C at a low voltage (10 V), which indicates the promise of its heat transfer properties for anti-icing applications.

3. When comprehensively considering the hydrophobicity, mechanical strength characteristics, de-icing effect, and economic cost of the concrete filled with CNF/GLC at a 1.0 wt.% rate, it is clear that it achieves the best results for practical applications. In other words, the carbonaceous additive reduces the cost of electrically conductive concrete by 43% compared to the same cost for nano-sized carbon (at 6 wt.% loading), and by 37% compared to that of recycled metal shavings (at 1.5 wt.% loading).

The results show that the CNF/GLC composite has good electrical resistance-heating efficiency. It can be applied in the de-icing of concrete pavement to protect the environment, save natural resources, and utilize agricultural waste. In the future, concrete filled with CNF/GLC would be an attractive candidate for pavement in climatic areas subject to sub-zero temperatures. We propose that future works should be focused on the long-term investigation of the electroconductivity and advanced study of the mechanical properties of concrete composites with CNF/GLC and other types of conductive carbon as additives at various ratios.

Author Contributions: C.D.: Conceptualization, Writing—original draft; F.S., Z.M.: Project administration, Writing—original draft; A.V.K., M.Y.: Methodology; A.T. (Azamat Taurbekov), B.B., A.T. (Aidos Tolynbekov): Investigation, Writing—Review & Editing. All authors have read and agreed to the published version of the manuscript.

Funding: This research is funded by the Science Committee of the Ministry of Science and Higher Education of the Republic of Kazakhstan (Grant No. OR11465430).

Data Availability Statement: Data sharing is not applicable to this article.

Conflicts of Interest: The authors declare no conflict of interest.

References

1. Chen, L.; Yang, Q.; Yang, X.; Liu, Z.; Song, Q. Numerical Investigation of Heat Transfer Performance of Graphene-Doped Anti-/Deicing Component. *Therm. Sci. Eng. Prog.* **2022**, *28*, 101098. [CrossRef]
2. Chen, L.; Zhang, Y.; Liu, Z.; Song, Q.; Liu, C. Effect of Graphene Oxide Doping on Anti-/Deicing Performance of Shape Memory Epoxy Resin. *Mater. Today Commun.* **2022**, *30*, 103025. [CrossRef]
3. Wang, B.; Yu, P.; Yang, Q.; Jing, Z.; Wang, W.; Li, P.; Tong, X.; Lin, F.; Wang, D.; Lio, G.E.; et al. Upcycling of Biomass Waste into Photothermal Superhydrophobic Coating for Efficient Anti-Icing and Deicing. *Mater. Today Phys.* **2022**, *24*, 100683. [CrossRef]
4. Kenzhebayeva, A.; Bakbolat, B.; Sultanov, F.; Daulbayev, C.; Mansurov, Z. A Mini-Review on Recent Developments in Anti-Icing Methods. *Polymers* **2021**, *13*, 4149. [CrossRef]
5. Bakr, A.R.; Fu, G.Y.; Hedeen, D. Water Quality Impacts of Bridge Stormwater Runoff from Scupper Drains on Receiving Waters: A Review. *Sci. Total Environ.* **2020**, *726*, 138068. [CrossRef]
6. Novotny, E.V.; Murphy, D.; Stefan, H.G. Increase of Urban Lake Salinity by Road Deicing Salt. *Sci. Total Environ.* **2008**, *406*, 131–144. [CrossRef]
7. Sanzo, D.; Hecnar, S.J. Effects of Road De-Icing Salt (NaCl) on Larval Wood Frogs (Rana Sylvatica). *Environ. Pollut.* **2006**, *140*, 247–256. [CrossRef]
8. Kayama, M.; Quoreshi, A.M.; Kitaoka, S.; Kitahashi, Y.; Sakamoto, Y.; Maruyama, Y.; Kitao, M.; Koike, T. Effects of Deicing Salt on the Vitality and Health of Two Spruce Species, Picea Abies Karst., and Picea Glehnii Masters Planted along Roadsides in Northern Japan. *Environ. Pollut.* **2003**, *124*, 127–137. [CrossRef]
9. Rivett, M.O.; Cuthbert, M.O.; Gamble, R.; Connon, L.E.; Pearson, A.; Shepley, M.G.; Davis, J. Highway Deicing Salt Dynamic Runoff to Surface Water and Subsequent Infiltration to Groundwater during Severe UK Winters. *Sci. Total Environ.* **2016**, *565*, 324–338. [CrossRef]
10. Alamayreh, M.I.; Alahmer, A.; Younes, M.B.; Bazlamit, S.M. Pre-Cooling Concrete System in Massive Concrete Production: Energy Analysis and Refrigerant Replacement. *Energies* **2022**, *15*, 1129. [CrossRef]
11. Li, P.; Li, W.; Sun, Z.; Shen, L.; Sheng, D. Development of Sustainable Concrete Incorporating Seawater: A Critical Review on Cement Hydration, Microstructure and Mechanical Strength. *Cem. Concr. Compos.* **2021**, *121*, 104100. [CrossRef]
12. Shishegaran, A.; Daneshpajoh, F.; Taghavizade, H.; Mirvalad, S. Developing Conductive Concrete Containing Wire Rope and Steel Powder Wastes for Route Deicing. *Constr. Build. Mater.* **2020**, *232*, 117184. [CrossRef]

13. Gao, J.; Guo, H.; Wang, X.; Wang, P.; Wei, Y.; Wang, Z.; Huang, Y.; Yang, B. Microwave Deicing for Asphalt Mixture Containing Steel Wool Fibers. *J. Clean. Prod.* **2019**, *206*, 1110–1122. [CrossRef]
14. Dong, W.; Huang, Y.; Lehane, B.; Aslani, F.; Ma, G. Mechanical and Electrical Properties of Concrete Incorporating an Iron-Particle Contained Nano-Graphite by-Product. *Constr. Build. Mater.* **2021**, *270*, 121377. [CrossRef]
15. Chen, L.; Ren, J.; Zhang, Y.; Liu, Z.; Xu, F.; Cheng, J. Microstructure Simulation and Heat Transfer Optimization for Graphene Oxide Doped Anti-/Deicing Composites. *Compos. Sci. Technol.* **2021**, *213*, 108901. [CrossRef]
16. Sultanov, F.; Bakbolat, B.; Daulbaev, C.; Urazgalieva, A.; Azizov, Z.; Mansurov, Z.; Tulepov, M.; Pei, S.S. Sorptive Activity and Hydrophobic Behavior of Aerogels Based on Reduced Graphene Oxide and Carbon Nanotubes. *J. Eng. Phys. Thermophys.* **2017**, *90*, 826–830. [CrossRef]
17. Lu, S.; Bai, E.; Xu, J.; Chen, J. Research on Electromagnetic Properties and Microwave Deicing Performance of Carbon Fiber Modified Concrete. *Constr. Build. Mater.* **2021**, *286*, 122868. [CrossRef]
18. Liu, Z.; Yang, X.; Wang, Y.; Luo, S. Engineering Properties and Microwave Heating Induced Ice-Melting Performance of Asphalt Mixture with Activated Carbon Powder Filler. *Constr. Build. Mater.* **2019**, *197*, 50–62. [CrossRef]
19. Atamanov, M.; Amrousse, R.; Jandosov, J.; Hori, K.; Kerimkulova, A.; Chenchik, D.; Kolesnikov, B. Combustion Characteristics of HAN-Based Green Propellant Assisted with Nanoporous Active Carbons. *Eurasian Chem.-Technol. J.* **2017**, *19*, 215. [CrossRef]
20. Daulbayev, C.; Sultanov, F.; Korobeinyk, A.V.; Yeleuov, M.; Taurbekov, A.; Bakbolat, B.; Umirzakov, A.; Baimenov, A.; Daulbayev, O. Effect of Graphene Oxide/Hydroxyapatite Nanocomposite on Osteogenic Differentiation and Antimicrobial Activity. *Surf. Interfaces* **2022**, *28*, 101683. [CrossRef]
21. Yeleuov, M.; Seidl, C.; Temirgaliyeva, T.; Taurbekov, A.; Prikhodko, N.; Lesbayev, B.; Sultanov, F.; Daulbayev, C.; Kumekov, S. Modified Activated Graphene-Based Carbon Electrodes from Rice Husk for Supercapacitor Applications. *Energies* **2020**, *13*, 4943. [CrossRef]
22. Yeleuov, M.; Daulbayev, C.; Taurbekov, A.; Abdisattar, A.; Ebrahim, R.; Kumekov, S.; Prikhodko, N.; Lesbayev, B.; Batyrzhan, K. Synthesis of Graphene-like Porous Carbon from Biomass for Electrochemical Energy Storage Applications. *Diam. Relat. Mater.* **2021**, *119*, 108560. [CrossRef]
23. Ming, Y.; Duan, Y.; Zhang, S.; Zhu, Y.; Wang, B. Self-Heating 3D Printed Continuous Carbon Fiber/Epoxy Mesh and Its Application in Wind Turbine Deicing. *Polym. Test.* **2020**, *82*, 106309. [CrossRef]
24. Pan, L.; Liu, Z.; kızıltaş, O.; Zhong, L.; Pang, X.; Wang, F.; Zhu, Y.; Ma, W.; Lv, Y. Carbon Fiber/Poly Ether Ether Ketone Composites Modified with Graphene for Electro-Thermal Deicing Applications. *Compos. Sci. Technol.* **2020**, *192*, 108117. [CrossRef]
25. Sassani, A.; Arabzadeh, A.; Ceylan, H.; Kim, S.; Sadati, S.M.S.; Gopalakrishnan, K.; Taylor, P.C.; Abdualla, H. Carbon Fiber-Based Electrically Conductive Concrete for Salt-Free Deicing of Pavements. *J. Clean. Prod.* **2018**, *203*, 799–809. [CrossRef]
26. Rhee, I.; Lee, J.-S.; Kim, J.H.; Kim, Y.A. Thermal Performance, Freeze-and-Thaw Resistance, and Bond Strength of Cement Mortar Using Rice Husk-Derived Graphene. *Constr. Build. Mater.* **2017**, *146*, 350–359. [CrossRef]
27. Daulbayev, C.; Sultanov, F.; Korobeinyk, A.V.; Yeleuov, M.; Azat, S.; Bakbolat, B.; Umirzakov, A.; Mansurov, Z. Bio-Waste-Derived Few-Layered Graphene/SrTiO3/PAN as Efficient Photocatalytic System for Water Splitting. *Appl. Surf. Sci.* **2021**, *549*, 149176. [CrossRef]
28. Amiri, M.; Golmohammadi, F. Biomass Derived Hierarchical 3D Graphene Framework for High Performance Energy Storage Devices. *J. Electroanal. Chem.* **2019**, *849*, 113388. [CrossRef]
29. Singh, N.; Sharma, M.; Mondal, D.; Maru, D.A.; Rathod, M.R.; Sequeira, R.A.; Chudasama, N.A.; Prasad, K. Seaweed Biomass Derived Bio Solvents for the Large Scale Production of Few Layered Graphene Nanosheets from Graphite. *Mater. Sci. Energy Technol.* **2021**, *4*, 100–106. [CrossRef]
30. Prikhod'ko, N.G.; Mansurov, Z.A.; Auelkhankyzy, M.; Lesbaev, B.T.; Nazhipkyzy, M.; Smagulova, G.T. Flame Synthesis of Graphene Layers at Low Pressure. *Russ. J. Phys. Chem. B* **2015**, *9*, 743–747. [CrossRef]
31. Ferrari, A.C. Raman Spectroscopy of Graphene and Graphite: Disorder, Electron–Phonon Coupling, Doping and Nonadiabatic Effects. *Solid State Commun.* **2007**, *143*, 47–57. [CrossRef]
32. Stobinski, L.; Lesiak, B.; Malolepszy, A.; Mazurkiewicz, M.; Mierzwa, B.; Zemek, J.; Jiricek, P.; Bieloshapka, I. Graphene Oxide and Reduced Graphene Oxide Studied by the XRD, TEM and Electron Spectroscopy Methods. *J. Electron. Spectrosc. Relat. Phenom.* **2014**, *195*, 145–154. [CrossRef]
33. Wu, S.; Mo, L.; Shui, Z.; Chen, Z. Investigation of the Conductivity of Asphalt Concrete Containing Conductive Fillers. *Carbon* **2005**, *43*, 1358–1363. [CrossRef]
34. AlShareedah, O.; Nassiri, S. Spherical Discrete Element Model for Estimating the Hydraulic Conductivity and Pore Clogging of Pervious Concrete. *Constr. Build. Mater.* **2021**, *305*, 124749. [CrossRef]
35. Shen, L.; Yao, X.; Zhu, D.; Alkayem, N.F.; Cao, M.; Ren, Q. A Thermal Cracking Pattern-Based Multiscale Homogenization Method for Effective Thermal Conductivity of Steel Fiber Reinforced Concrete after High Temperature. *Int. J. Heat Mass Transf.* **2021**, *180*, 121732. [CrossRef]
36. Dehghanpour, H.; Yilmaz, K.; Ipek, M. Evaluation of Recycled Nano Carbon Black and Waste Erosion Wires in Electrically Conductive Concretes. *Constr. Build. Mater.* **2019**, *221*, 109–121. [CrossRef]
37. Zhu, X.; Bai, Y.; Chen, X.; Tian, Z.; Ning, Y. Evaluation and Prediction on Abrasion Resistance of Hydraulic Concrete after Exposure to Different Freeze-Thaw Cycles. *Constr. Build. Mater.* **2022**, *316*, 126055. [CrossRef]

38. Saidmurodov, S.S.; Jacobsen, S.; Hendriks, M.A.N.; Shamsutdinova, G. An Evaluation of the Ice Melting during Concrete-Ice Abrasion Experiment. *Case Stud. Therm. Eng.* **2022**, *35*, 102088. [CrossRef]
39. Wang, C.; Fu, H.; Ma, W.; Zhang, Z.; Ji, X.; Han, X. Combination Design and Performance Evaluation of Conductive Bonding Layer for Asphalt Pavement Active Deicing. *Constr. Build. Mater.* **2020**, *263*, 121037. [CrossRef]
40. Liu, X.; Zhao, Y.; Liu, W.; Yan, D. Microwave Heating and Deicing Efficiency for Asphalt Concrete with SiC–Fe3O4 Microwave Enhanced Functional Layer. *J. Clean. Prod.* **2022**, *332*, 130111. [CrossRef]
41. Rao, R.; Fu, J.; Chan, Y.; Tuan, C.Y.; Liu, C. Steel Fiber Confined Graphite Concrete for Pavement Deicing. *Compos. Part B Eng.* **2018**, *155*, 187–196. [CrossRef]
42. Zhao, H.; Wu, Z.; Wang, S.; Zheng, J.; Che, G. Concrete Pavement Deicing with Carbon Fiber Heating Wires. *Cold Reg. Sci. Technol.* **2011**, *65*, 413–420. [CrossRef]
43. Liu, J.; Xu, J.; Huang, H.; Chen, H. Microwave Deicing Efficiency and Dielectric Property of Road Concrete Modified Using Different Wave Absorbing Material. *Cold Reg. Sci. Technol.* **2020**, *174*, 103064. [CrossRef]

coatings

MDPI

Article

Development of a Novel Double-Sulfate Composite Early Strength Agent to Improve the Hydration Hardening Properties of Portland Cement Paste

Yuli Wang , Luyi Sun, Songhui Liu *, Shuaijie Li, Xuemao Guan and Shuqiong Luo

Henan Key Laboratory of Materials on Deep-Earth Engineering, School of Materials Science and Engineering, Henan Polytechnic University, Jiaozuo 454003, China

* Correspondence: liusonghui@hpu.edu.cn

Abstract: A novel double-sulfate composite early strength agent (DSA) incorporating aluminum sulfate and sodium sulfate was developed to improve the early strength of Portland cement paste. The effect of the DSA dosage on the setting and hardening properties, hydration process, hydration product composition, microstructure, and pore structure of the Portland cement paste was investigated to reveal its synergistic enhancement mechanism. The results show that the 3 d and 28 d compressive strengths of the Portland cement paste incorporating with 1.0% aluminum sulfate and 1.5% sodium sulfate performed the best, with a 21.3% and 29.7% increase, respectively, compared to the control group. The heat of hydration, XRD, TG, SEM, and MIP tests showed that aluminum sulfate and sodium sulfate acted synergistically, with more AFt (Ettringite) being produced by the synergistic use of 1.0% aluminum sulfate and 1.5% sodium sulfate. Moreover, the hydration of C_3S and C_2S was accelerated, which resulted in a denser microstructure.

Keywords: Portland cement; aluminum sulfate; sodium sulfate; hydration; synergistic enhancement mechanism

1. Introduction

Nowadays, silicate cement, also known as Portland cement, is an indispensable material for the construction industry. With the continuous development of society and the continuous innovation of infrastructure today, the demand for cement is huge. Yet, to satisfy specific requirements and save time, admixtures are generally required to increase the early strength of Portland cement [1–5]. For example, Portland cement with water-reducing and early-strengthening agents is often used when carrying out the projects for precast elements, construction of paving, grouting, and filling [6], which can significantly speed up the process of the project. At present, there are three main types of early strength agents being used: organic salts, inorganic salts, and compounds [7,8]. In addition, some special materials can also improve the strength of cement paste [9–12]. Of these, inorganic salts are widely used as their effect on Portland cement is often immediate and obvious [13]. However, the use of an early-strengthening agent containing only one inorganic salt for cement has certain limitations (for example, the use of aluminum sulfate alone does not result in a cement that exhibits higher early strength) [14–16], so the study of composite early-strengthening agents that can better improve the performance of cement is the primary trend of current research [3,17].

Aluminum sulfate is widely used as an alkali-free early strengthener with low contamination and good stability, and it improves worker safety [18,19]. Research shows that aluminum sulfate added to cement promotes hydration and hardening [3,20–22]. Chen C et al. [22] supported that aluminum sulfate dissociates aluminum ions and sulfate ions in the cement paste, which will combine with calcium and hydroxide ions in the cement to form AFt (Ettringite). The exothermic generation of AFt from the above reactions promotes the hydration of C_3A and the production of more AFt, resulting in reduced flow and setting time of the cement paste. Guosheng Ren et al. [21] found that the addition of aluminum sulfate resulted in the rapid generation of AFt during the initial reaction period and advanced the hydration of C_3S. The high-valent metal ion Al^{3+} can squeeze the diffusion bilayer of C–S–H gels, which promotes the agglomeration of C–S–H gels [3]. The early strength agent studied by Yang Renhe et al. [20] can make cement paste set and strengthen early, the main component of this early strength agent is aluminum sulfate.

The use of sodium sulfate early strengthening agents to modify cement properties has also been studied [2,13,23–26]. It has been found that when sodium sulfate is added to cement, the introduced sulfate and the calcium ions in the cement will participate in the hydration of C_3A and promote the production of AFt [23,27]. The combination of sodium and hydroxide ions leads to an increase in the alkalinity of Portland cement paste, which increases the solubility of C_3A and calcium sulfate and promotes the hydration of C_3S [2,13]. However, the soluble salts formed by the sodium ions can dissolve and precipitate, causing swelling and cracking of the cement [13,28].

In conclusion, the use of a certain amount of aluminum sulfate or sodium sulfate alone will result in cement exhibiting higher early strengths. However, it is less clear whether the synergistic use of sodium sulfate and aluminum sulfate will result in cement with more desirable early strength and compensate for the later strength reversal caused by the addition of sodium sulfate. In addition, the doping ratios for synergistic use and the mechanism of action during synergistic use have not been well understood.

To address these issues, this work investigated the effect of different ratios of aluminum sulfate and sodium sulfate on the performance of the cement and investigated the appropriate synergistic balance to achieve a better early strength of the cement, as well as an excellent overall performance. The synergistic mechanism of aluminum sulfate and sodium sulfate on the hydration and hardening of the cement was also investigated by the heat of hydration, XRD, TG, SEM, and MIP.

2. Experimental Materials and Experimental Protocol

2.1. Experimental Materials

In this experiment, the Ordinary Portland cement (OPC) of PO·42.5 was used, the properties of which meet the standard Chinese GB175-2007 [29], the specific composition and properties of this cement are shown in Tables 1–3 below. Aluminum sulfate ($Al_2(SO_4)_3 \cdot 18H_2O$) and anhydrous sodium sulfate (Na_2SO_4) are chemically pure reagents, and the purity is more significant than 99.0%. The water used is laboratory tap water.

Table 1. The physical properties of the Portland cement.

Fineness/%	SSA (m^2/kg)	Stability	Setting Time /Min		Flexural Strength/MPa		Compressive /MPa	
			Initial	Final	3 d	28 d	3 d	28 d
0.8	366	qualified	200	250	6.0	9.5	30.1	50.5

Table 2. The chemical composition (wt%) of the Portland cement.

CaO	SiO_2	Al_2O_3	Fe_2O_3	MgO	SO_3	Na_2O	K_2O	Loss
61.86	23.17	5.37	3.32	2.78	2.36	0.36	0.20	0.13

Table 3. The mineral composition (wt%) of the Portland cement.

C_3S	C_2S	C_3A	C_4AF	Gypsum	Calcite	Amorphous
58.68	15.03	9.34	7.80	2.65	3.62	2.88

2.2. *Experimental Protocol*

2.2.1. Design Ratios

Table 4 lists the design groups of this experiment. The water to cement ratio in this experiment was 0.4, and the chemical admixture was sieved through 200 mesh and used.

Table 4. Design ratio.

Groups	$Al_2(SO_4)_3 \cdot 18H_2O$ [1]/%	Na_2SO_4 [1]/%	Cement/g	Water/g
Control	-	-	500	200
Al0.5%	0.5	-	500	200
Al1.0%	1.0	-	500	200
Al1.5%	1.5	-	500	200
Al2.0%	2.0	-	500	200
Na0.5%	-	0.5	500	200
Na1.0%	-	1.0	500	200
Na1.5%	-	1.5	500	200
Na2.2%	-	2.0	500	200
Al1.0% + Na0.5%	1.0	0.5	500	200
Al1.0% + Na1.0%	1.0	1.0	500	200
Al1.0% + Na1.5%	1.0	1.5	500	200
Al1.05 + Na2.0%	1.0	2.0	500	200

[1] mixed in as a percentage by mass of cement.

2.2.2. Preparation of Portland Cement Paste

The prepared aluminum sulfate and sodium sulfate was ground to a powder and then mixed with cement and stirred well. According to the General "Portland Cement" National Standard (GB175-2007) [29], it takes 15 s to add the mixed cement and water to the mixing pot, then stirs for 2 min at low speed, then shakes and finally stir for 2 min at high speed.

2.2.3. Test for Setting Time

The Vicar method was used to test the setting time according to the "Test Method for the Standard Consistency, Setting Time and Settlement of Cement" (GB/T1346-2011) [30]. The setting time was tested separately three times to obtain the average.

2.2.4. Test for Compressive Strength

The prepared paste was molded in a clean 40 × 40 × 40 mm mold and covered with cling film to prevent water evaporation, then placed in a standard curing room for one day, after which de-molded and continue curing was conducted. The standard curing room temperature was 20 ± 2 °C and the humidity was ≥95%, in line with the curing conditions.

According to the "Standard for Mechanical Properties of Plain Concrete Test Methods" (GB50081-2019) [31], compressive strength tests were carried out on 3 d and 28 d specimens, with each group of specimens measured three times and the average value taken.

2.2.5. Hydration Heat Release

The hydration heat of Portland cement with different admixtures was carried by using an isothermal conductivity calorimeter (ICC, TAM Air, Newcastle, DE, USA). A sample of 2.5 g of cement mixture according to Table 4 was blended into the calorimetric apparatus and equilibrated for two hours to exclude heat flow errors after installation. For each group, the mixture test at 20 °C.

2.2.6. XRD Analysis

The specimens that had been soaked in anhydrous ethanol for 24 h were dried in a vacuum oven at 60 °C, then ground into a powder and sieved through 200 mesh. A Smart lab X-ray diffractometer (CuKα radiation, λ = 1.54056 Å, Rigaku Smart Lab, Tokyo, Japan) at a scan rate of 10 °C/min and an angle of 5–80° was used for the XRD analysis.

2.2.7. TGA Analysis

The TGA tests were started on the dried and ground powder using an HCT-3 microcomputer differential thermal equilibrium (Rigaku, Tokyo, Japan) with a carry temperature of 40–900 °C and a ramp-up rate of 10 °C per minute. Samples from each group of cement, after curing for 3 days and 28 days, were used for testing. After curing, the main composition of the medium (AFt, $Ca(OH)_2$ and $CaCO_3$) in the cement paste was analyzed. According to the study, the mass percentage of AFt produced (m_{AFt}) was calculated based on the weight loss from dehydration of AFt at 50~150 °C (ignoring the loss of other components bound to water). The mass percentage of calcium hydroxide produced (m_{CH}) was calculated from the weight loss when CH was dehydrated at 400~500 °C and $CaCO_3$ was decomposed at 550~900 °C [32,33]. The following are the relevant chemical reactions.

$$3CaO\cdot Al_2O_3\cdot 3CaSO_4\cdot 32H_2O \rightarrow 3CaO\cdot Al_2O_3\cdot 3CaSO_4 + 32H_20 \tag{1}$$

$$Ca(OH)_2 \rightarrow CaO + H_2O\uparrow \tag{2}$$

$$CaCO_3 \rightarrow CaO + CO_2\uparrow \tag{3}$$

The following are the content calculations.

$$m_{AFt}(\%) = m_{H_2O} \times \frac{M_{AFt}}{32M_H} \tag{4}$$

$$m_{CH}(\%) = m_H \times \frac{M_{CH}}{M_H} + m_{CC} \times \frac{M_{CC}}{M_{CO_2}} \tag{5}$$

where, m_{AFt} is the mass fraction percentage of AFt produced, m_{H_20} is the mass fraction percentage of water loss from crystallization lost by thermal decomposition of the thermal decomposition of AFt, M_{AFt} is the molar mass fraction of AFt, m_H is the mass fraction percentage of water loss by decomposition of CH, m_{CC} is the mass fraction percentage of weight loss by decomposition of $CaCO_3$, M_H indicates the molar mass of H_2O, M_{CC} is the molar mass of $CaCO_3$ and M_{CO_2} is the molar mass of CO_2.

2.2.8. SEM Analysis

According to the sample-making requirements, the middle part of the specimen was taken separately and sealed in anhydrous ethanol for 24 h to terminate the hydration. Before starting the SEM experiment, the samples were dried in a vacuum oven at 40 °C for a further 24 h, then the samples were treated for electrical conductivity and tested with a Merlin Compact SEM scanner (Carl Zeiss NTS GmbH, Oberkohan, Germany).

2.2.9. MIP Analysis

The specimens with ages of 28 d were cut to samples of 10 mm in height and width and 15 mm in length, then they were sealed in anhydrous ethanol for 24 h to terminate hydration, and then put into a vacuum drying oven to dry at 40 °C for 24 h. The porosity and pore size distribution of the samples was tested by using the Automatic Pores IV 9500 Mercury Porosimeter (Micromeritics, Atlanta, GA, USA).

3. Results and Discussion

3.1. *Effect of the DSA on the Setting Time of Portland Cement Paste*

The effect of the DSA on the setting time of Portland cement paste is shown in Figure 1. It shows that the initial setting time and final setting time decreased continuously with the increase of aluminum sulfate dosing. Compared to the control group, for each increase of 0.5% aluminum sulfate, the initial setting time decreased by 24.0%, 31.6%, 41.3%, 47.4%, and the final setting time decreased by 19.6%, 26.4%, 28.4%, 35.2%, respectively. The effect of sodium sulfate on the initial and final setting times of Portland cement paste was less pronounced than that of aluminum sulfate; the initial and final setting times of Portland cement paste fluctuated upwards and downwards as the amount of sodium sulfate was increased. Compared to the control group, for each increase of 0.5% sodium sulfate, the initial setting time of the cement was reduced by −4.6%, 3.6%, 9.2%, 5.6%, and the final setting time by 4.4%, −2.4%, 0.8%, 8.4%, respectively. When the DSAs were used in combination, the initial and final setting times of the cement were insignificant when the amount of sodium sulfate was increased by 0.5% and using 1.0% aluminum sulfate as a quantification. Compared to the control group, the initial setting time was reduced by 37.2%, 38.3%, 30.1%, 28.6%, and the final setting time was reduced by 26.8%, 16.0%, 24.4%, 24.0%, respectively, while compared to the Al1.0% group, the initial setting time was reduced by 8.21%, 9.70%, −2.24%, −4.48%, and the final setting time was reduced by 0.54%, −14.13%, −2.72%, −3.26%, respectively. In summary, the setting time of the Portland cement paste decreases as the aluminum sulfate dose increases; in contrast, the effect of sodium sulfate on the setting time of the Portland cement paste is not significant. When aluminum sulfate and sodium sulfate were used in combination, it was the aluminum sulfate that had the main effect on the setting time of the Portland cement paste.

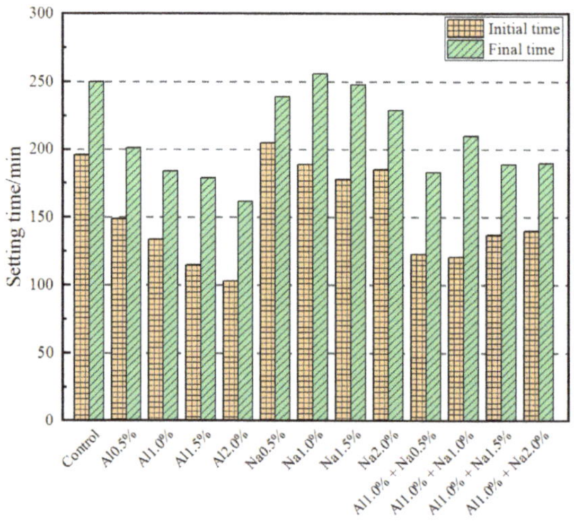

Figure 1. Effect of different amounts of the DSA on setting time of Portland cement paste.

3.2. *Effect of the DSA on the Compressive Strength of Portland Cement Paste*

The compressive strengths of the Portland cement paste at 3 d and 28 d for different doses of the DSA are shown in Figure 2. It was observed that the compressive strength of Portland cement paste at 3 d and 28 d increased first and then decreased with increasing admixture levels, whether aluminum sulfate or sodium sulfate was used alone or in combination with aluminum sulfate and sodium sulfate. Compared to the control group, for each 0.5% increase in aluminum sulfate, the compressive strength of Portland cement paste at 3 d was enhanced by 1.8%, 7.2%, 6.0%, 4.8%, and the compressive strength at 28 d by 7.9%, 22.4%, 19.4%, −4.4%, respectively. The best early strength was achieved at 1.0% of aluminum sulfate and no inversion of strength occurred at a later stage. Compared to the control group, for each 0.5% increase in sodium sulfate, the compressive strength of Portland cement paste at 3 d was increased by 9.6%, 10.2%, 10.5%, 0.0%, and the compressive strength at 28 d was increased by 6.7%, 7.9%, 20.6%, −14.9% for each 0.5% increase in sodium sulfate. Although the compressive strengths of the Na1.0% and Na1.5% groups at 3 d were almost the same, the compressive strength of the Na1.5% group at 28 d was much greater than that of the Na1.0% group, so the 1.5% sodium sulfate admixture had the best effect in improving the compressive strength of the cement. The results show that 1.0% aluminum sulfate and 1.5% sodium sulfate have the greatest influence on the compressive strength of Portland cement paste, the strength at 3 d was enhanced by 13.2% and 10.1% over the Al1.0% and Na1.5% groups, respectively, and the strength at 28 d was enhanced by 29.7%, 5.9% and 7.5% over the control, Al1.0%, and Na1.5% groups, respectively, with no decrease in compressive strength compared to the blank group.

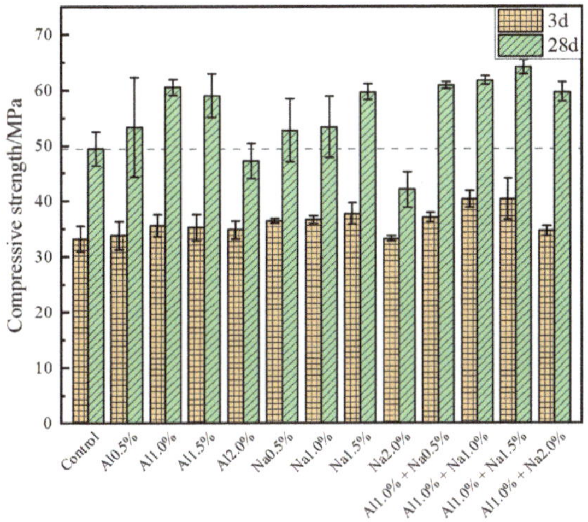

Figure 2. Effect of different amounts of the DSA on compressive strength of Portland cement.

3.3. Early Hydration Heat Evolution of Portland Cement Paste

The hydration exothermic rates and accumulative hydration heats of Portland cement paste for the control, Al1.0%, Na1.5%, and Al1.0% + Na1.5% groups are shown in Figure 3a,b. The early hydration of Portland cement was divided into five main periods, as shown in Figure 3a,b for I, II, III, IV and V, which indicated the initial reaction period (0–0.8 h), the induction period (0.8–3.6 h), the acceleration period (3.6–12.5 h), the deceleration period (12.5–30 h), and the stabilization period (>30 h), respectively [21,33,34].

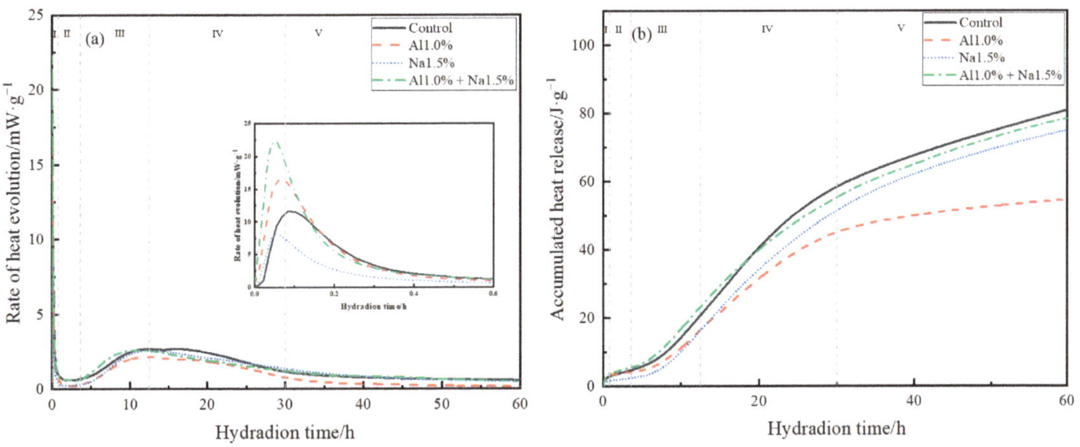

Figure 3. (**a**) hydration exothermic rates and (**b**) accumulative hydration heats of Portland cement paste.

As shown in Figure 3a, during the induction period, compared to the control group, the Al1.0% group accelerated hydration and increased the maximum rate of hydration during the induction period, while the Na1.5% group also promoted hydration at the beginning of the induction period and brought the peak earlier, but reduced the maximum rate of hydration during the induction period. The Al1.0% + Na1.5% group greatly facilitated the hydration of the cement during the induction period, with the highest hydration rate during the induction period being approximately twice that of the control group. According to previous studies, the hydration of C_3A to produce AFt and the reaction of aluminum sulfate and calcium hydroxide to produce AFt mainly take place during the induction period [35], therefore, it is believed that the Al1.0% + Na1.5% group greatly promoted the hydration of the cement during the induction period, with a large amount of AFt being produced. The hydration of C_3S occurred mainly during the acceleration, deceleration, and stabilization periods [21]. It was observed that the Al1.0% + Na1.5% group caused the maximum hydration rate to occur earlier, thus promoting the hydration of C_3S during the acceleration period and that there was also a very weak promotion of C_3S hydration after 30 h.

According to Figure 3b, the cumulative hydration exotherm of the Al1.0% + Na1.5% group was higher than the other groups until about 20 h, and the cumulative hydration exotherm was smaller than that of the control group after 20 h hours. According to previous studies, the main factor affecting the rate of hydration of cement during the deceleration and stabilization phases is the diffusion rate [23], so the main reason for this phenomenon in the Al1.0% + Na1.5% group may be that the cement had formed a denser structure at this point compared to control, weakening the diffusion rate.

3.4. Hydration Products Analysis

3.4.1. XRD Analysis

As shown in Figure 4, XRD analysis of specimens at 3 d and 28 d for the Al1.0%, Na1.5% and Al1.0% + Na1.5% groups revealed that the main hydration products of Portland cement paste contained AFt, AFm (monosulfate), Ca $(OH)_2$, C–S–H, and un-hydrated products, such as C_3S, C_2S, C_4AF, and $CaCO_3$ that had been carbonated.

As shown in Figure 4a,b, with 3 d of hydration, the height of the AFt diffraction peak in the Al1.0% + Na1.5% group was slightly higher than the Al1.0% group, and much higher than the Na1.5% group and control group, indicating that the Al1.0% + Na1.5% group promoted the production of more AFt. Furthermore, looking at the calcium hydroxide diffraction peaks ($2\theta \approx 18°$), it is clear that their peak heights were in descending order for the Al1.0% + Na1.5% group, the Na1.5% group, the Al1.0% group and the control group. This indicates that the Al1.0% + Na1.5% group at 3d had the least amount of calcium hydroxide, which corresponds to the production of more AFt consuming more calcium ions. Although the control group had the lowest calcium hydroxide diffraction peak ($2\theta \approx 35°$) compared to the other groups, it had the strongest and highest calcium carbonate diffraction peak, probably due to the lower density of the control group, which absorbed more carbon dioxide and produced more calcium carbonate. The generated AFt is cross-linked to form the early skeleton in the hydration of Portland cement paste, which is filled with generated C–S–H, $CaCO_3$, etc. Then the generation of more AFt leads to a denser microstructure, which exhibits higher strength, and which confirms why the Al1.0% + Na1.5% group has a higher 3 d early strength.

According to Figure 4c,d, with 28 d of hydration, the AFt diffraction peaks of the Al1.0% + Na1.5% group are not very different from those of the Al1.0%, Na1.5% and control groups, but the AFm diffraction peaks are more pronounced than those of the other groups, which indicates that some of the AFt generated in the Al1.0% + Na1.5% group was converted to AFm. Comparing the diffraction peaks of calcium hydroxide ($2\theta \approx 18°$, $2\theta \approx 35°$) it is clear that the peak height of the Al1.0% + Na1.5% group was weaker while observing that the diffraction peaks of C_3S and C_2S were smaller for the Al1.0% + Na1.5% group than for the other three groups, which indicates that the synergistic use of 1.0% aluminum sulfate and 1.5% sodium sulfate promotes the hydration of Portland cement paste to a greater extent.

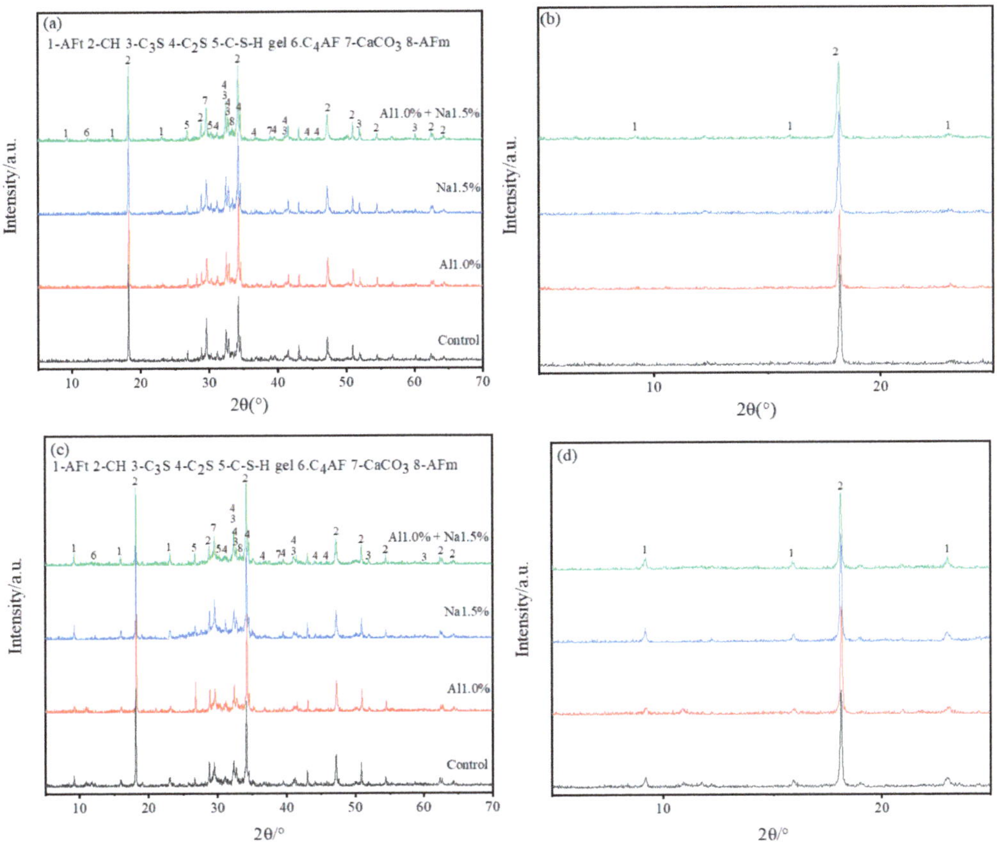

Figure 4. (**a–d**) are full-scale and locally magnified XRD patterns of Portland cement paste at 3 d and 28 d, respectively. (**a**) is full-scale XRD of Portland paste at 3 d, (**b**) is locally magnified XRD of Portland paste at 3 d, (**c**) is full-scale XRD of Portland paste at 28 d and (**d**) is locally magnified XRD of Portland paste at 28 d.

3.4.2. TG-DTG Analysis

To further analyze the hydration products composition of Portland cement paste, Figure 5a,b shows the TG-DTG curves of Portland cement paste at 3 d and 28 d. In the thermogravimetric analysis diagram of Portland cement paste, there are three main stages of heat absorption peaks, the first of which is subdivided into two subsections, heat loss from loss of bound water at 80–150 °C for AFt, heat loss from loss of bound water at 150–200 °C for AFm, heat loss from thermal decomposition of $Ca(OH)_2$ to CaO and water vapor at 370–480 °C, and heat loss from thermal decomposition of $CaCO_3$ to CaO and CO_2 at 490–800 °C [36–38].

According to Figure 5, the amount of AFt and CH produced can be calculated and shown in Table 5. Figure 5a and Table 5 shows that the amount of AFt generated in the Al1.0% group was more significant than that in the control group. While observing the DTG curves, it can found that the Al1.0% group had a larger peak at 150–200 °C and produced less CH than the control group, which indicates that the 1.0% of aluminum sulfate consumed some of the calcium ions and promoted the production of AFt, the reduction of CH and the hydration of C_3S, which in turn led to an increase in the early strength of Portland cement paste, as also proved by Liu et al. [3,21]. The AFt produced in the Na1.5% group was not significantly different compared to the control group, but the peak at 150–200 °C was slightly larger in the Na1.5% group than that in the control group, and the amount of CH produced in the Na1.5% group was much lower than in the control group, indicating that 1.0% of sodium sulfate promoted the hydration of C_3A and C_3S and improved the early strength of Portland cement paste, which is consistent with the studies of Mei et al. [23,33,39]. The Al1.0% + Na1.5% group had the highest AFt production. At the same time, a larger peak at 150–200 °C was observed in the DTG curve, which was due to a lower ratio of calcium sulfate to aluminate to promote conversion from more AFt into AFm [40–42]. as Additionally, there was a lower mass of CH production, which demonstrates that 1.0% of aluminum sulfate and 1.5% of sodium sulfate were synergistic, promoting more AFt generation, and better hydration of C_3A and C_3S, resulting in a tighter early structure and higher early strength at 3 d.

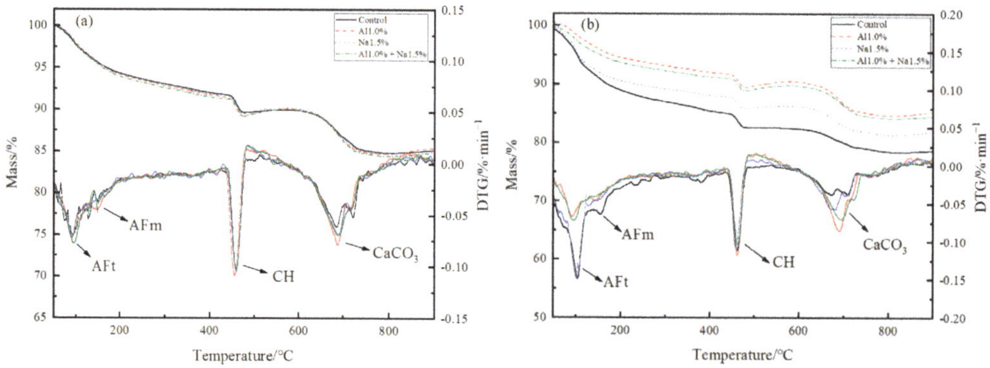

Figure 5. TG-DTG curves of Portland cement paste at (**a**) 3 d and (**b**) 28 d with different admixture.

Table 5. The amount of AFt and CH produced at 3 d by calculating the mass loss of Portland cement paste.

No.	m_{AFt}/%	m_H/%	m_{CC}/%	m_{CH}/%
Control	10.27	2.37	4.85	16.28
Al1.0%	10.81	2.02	4.86	15.53
Na1.5%	10.21	2.02	5.14	16.17
Al1.0% + Na1.5%	11.33	1.84	5.18	15.85

Figure 5b and Table 6 show that the CH production of the Al1.0% group was almost the same as that of the control, which indicated that 1.0% of aluminum sulfate promoted the hydration of C_3S. Brykov et al. [43] attributed the promotion of C_3S hydration by aluminum sulfate at a later stage to the compressive effect of Al^{3+} on the C–S–H layer, which promotes C–S–H aggregation. Higher CH production in the later stages caused more CH to be carbonated. He et al. also demonstrated that the $CaCO_3$ generated by carbonation filling the Portland cement paste pores could stabilize the Portland cement paste [44–46]. The Na1.5% group generated less CH than the control group. The phenomenon that the 1.5% of sodium sulfate promoted the hydration of C_3S, as shown by the heat of hydration and XRD analysis, as well as the increased compressive strength of the 28 d Na1.5% group, is thought to be the result of the addition of 1.5% of sodium sulfate promoting the generation of sodium hydroxide. The reduction of CH, along with the generation of sodium hydroxide enhances the solubility of C_3S, C_2S, which together promote the hydration of C_3S, C_2S [2,13,47]. According to Le Chatelier's principle [48,49], the CH production in the Al1.0% + Na1.5% group was less than the control and Al1.0% groups, but more than the Na1.5% group, so 1.0% of aluminum sulfate and 1.5% of sodium sulfate synergistically promote the hydration of late cement. Ratios of 1.0% of aluminum sulfate and 1.5% of sodium sulfate synergistically contribute more to the hydration of C_3S and C_2S in the later stages of Portland cement paste. Wang et al. [50] concluded that the development of late strength in cement was mainly related to the C–S–H generated by the hydration of C_3S and C_2S in cement. More C–S–H filling in the Portland cement paste aggregate leads to a denser structure and higher strength, which confirms that 1.0% of aluminum sulfate and 1.5% of sodium sulfate synergistically promote the hydration of C_3S and C_2S, with the best compressive strength performance at 28 d.

Table 6. The amount of AFt and CH produced at 28 d by calculating the mass loss of Portland cement paste.

No.	m_{AFt}/%	m_H/%	m_{CC}/%	m_{CH}/%
Control	18.96	3.33	3.89	16.24
Al1.0%	10.93	1.90	5.32	16.30
Na1.5%	17.45	2.22	4.44	15.02
Al1.0% + Na1.5%	11.15	1.75	5.31	15.95

3.5. Microstructure Observation

The SEM images (Figure 6a–h) of the 3 d Na1.5% group showed that a large amount of C–S–H was generated and covered the cement particles. At the same time, some CH and $CaCO_3$ were also filled in the cement skeleton, resulting in denser microstructure and better compressive strength than the control group at 3 d. As for the 3 d Al1.0% group, it clearly shows a large amount of AFt being densely interacted together, which makes the cement paste denser than the 3-d control group and has higher compressive strength. Observing the SEM of the Al1.0% + Na1.5% group reveals that there is dense AFt being intercalated, the amount of AFt is denser than that of the Al1.0% group and much denser than that of the control group. Moreover, it can see that more C–S–H was generated to be filled in compared to the other groups, which resulted in a denser structure and showed the desired early strength. These are consistent with the results of XRD and TG-DTG analyses that more AFt can be formed and the hydration of C_3S was also accelerated to make the Portland cement paste more compact.

Observation of the scanning electron microscopy pattern clearly shows that the Portland cement paste at 28 d was more fully hydrated than the Portland cement paste at 3 d. Although the Na1.5% group at 28 d also appears to have a dense structure, the structure is accompanied by large cracks, which affect the compressive strength of Portland cement paste. Cracking in the cement of the Na1.5% group at 28 d is thought to be initiated by water loss and re-crystallization of soluble sodium salts [13,28,51,52]. According to the observation of the mapping of the Al1.0% group at 28 d, it shows that the structure of this group was denser than that of the 28-d control group. According to the observation of the SEM pattern of the 28 d Al1.0% + Na1.5% group, it can be found that a large amount of C–S–H is filled in the cement paste. There were almost no large cracks on the surface of the 28 d Al1.0% + Na1.5% group compared to the 28-d control group, the 28 d Al1.0% group and the 28 d Na1.5% group. At the same time, there were some small particles embedded in the C–S–H, which would explain that structure of the 28 d Al1.0% + Na1.5% group was the densest; this phenomenon is better demonstrated in the chapter on the pore structure.

3.6. Pore Structure Evolution

The properties of an already hardened cement paste are related to the pore structure in addition to the already hydrated products and the anhydrate particles. The characterization of the pore structure involves porosity, pore size distribution, and pore size. The pore size in cementitious composites are classified into four categories according to the hazard posed by the size of the pores: no harmful pore size (<20 nm), less harmful pores (20–50 nm), harmful pores (50–200 nm), and more harmful pores (>200 nm) [53,54]. Relevant MIP test data are summarized in Table 7.

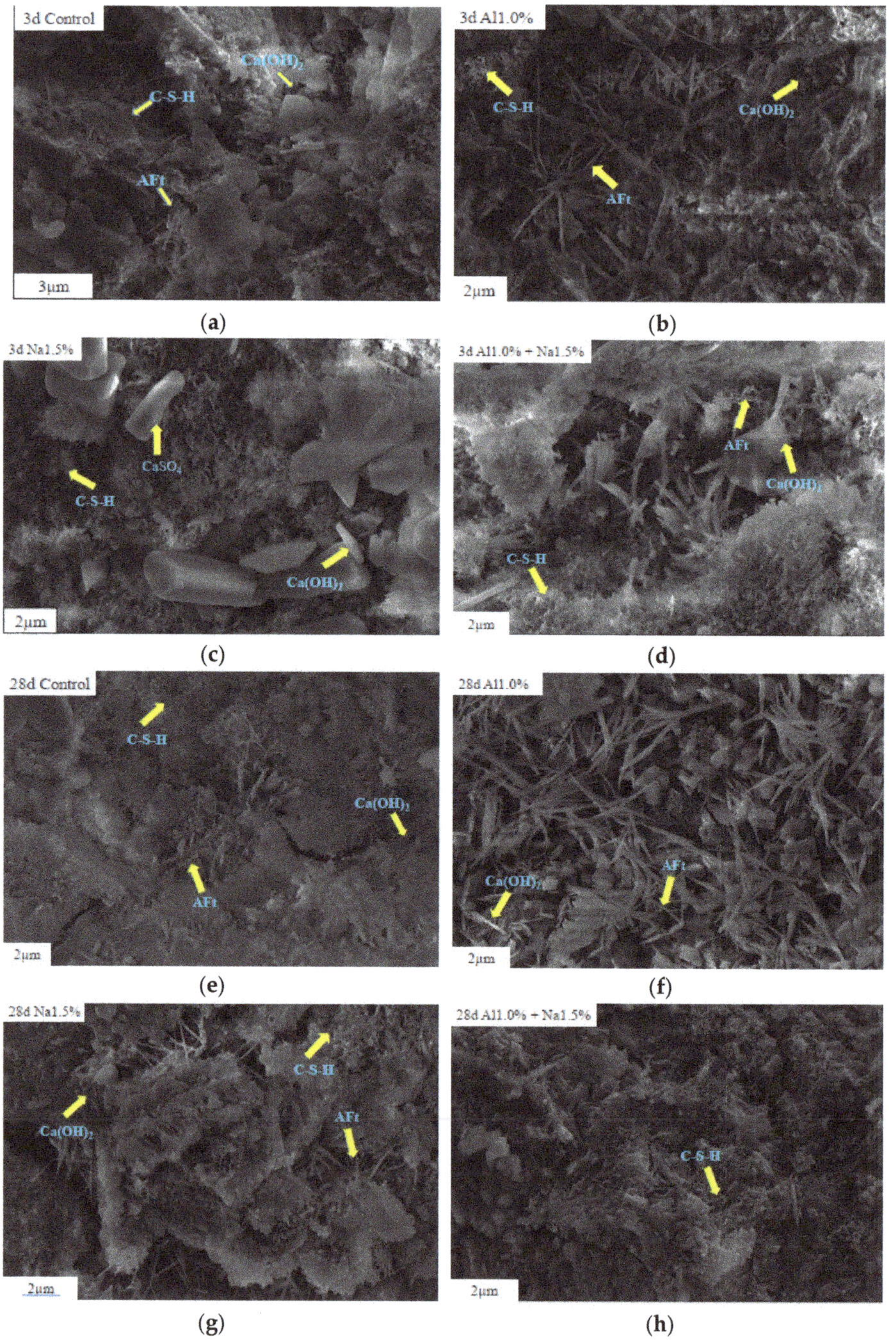

Figure 6. SEM images of Portland cement paste at 3 d and 28 d with different admixture. (**a**) is SEM images of Control at 3 d (**b**) is SEM images of Al1.0% at 3 d, (**c**) is SEM images of Na1.5% at 3 d, (**d**) is SEM images of Al1.0% + Na1.5% at 3 d, (**e**) is SEM images of Control at 28 d, (**f**) is SEM images of Al1.0% at 28 d, (**g**) is SEM images of Na1.5% at 28 d and (**h**) is SEM images of Al1.0% + Na1.5% at 28 d.

Table 7. Porosity and pore size distribution of paste with different admixture.

Groups	Average Pore Size (μm)	Porosity (%)	Proportion of Pore Size (%)			
			≤20 nm	20~50 nm	50~200 nm	≥200 nm
Control	0.03444	50.1	16.0	24.7	54.5	3.1
Al1.0%	0.03183	61.4	19.0	22.1	53.4	5.5
Na1.5%	0.03198	48.2	16.8	27.8	53.3	2.3
Al1.0% + Na1.5%	0.01736	36.1	45.7	36.3	11.9	5.8

According to Table 7, the Al1.0% + Na1.5%, Al1.0% and Na1.5% groups all reduced the average pore size of Portland cement paste compared to the control group, but the Al1.0% + Na1.5% group was the more effective. The average pore size for the Al1.0%+ Na1.5% group was 0.01736 μm, a reduction of 49.6% compared to the control group, while the Al1.0% and Na1.5% groups showed a decrease of 7.5% and 7.0%, respectively. Furthermore, according to Table 7, the Portland cement paste porosity was reduced in the Al1.0% + Na1.5% and Na1.5% groups compared to the control group. The total porosity of the Al1.0% + Na1.5% group was the smallest, with a 27.9% reduction compared to the control group. Initially, regarding the Al1.0% + Na1.5% group, no harm and less harmful pores accounted for 45.7% and 36.6% of the total porosity, respectively, while no harm and less harmful pores in the Na1.5% group accounted for 16.8% and 27.8% of the total porosity, respectively. Regarding the Al1.0% group, although the total porosity was higher than that of the control group, no harm and less harmful pores accounted for 19.0% and 22.1% of the total porosity, respectively, which was a higher percentage than the control group. Meanwhile according to Figure 7a,b, 1.0% aluminum sulfate and 1.5% of sodium sulfate affected the porosity and pore size distribution of Portland cement paste in varying degrees, which increases the compressive strength of Portland cement paste. The simultaneous use of 1.0% aluminum sulfate and 1.5% of sodium sulfate has a synergistic effect on the denseness of Portland cement paste, not only reducing the porosity but also decreasing the proportion of harmful and porous pores, which is one of the main reasons for the higher compressive strength at 28 d of the Al1.0% + Na1.5% group [55].

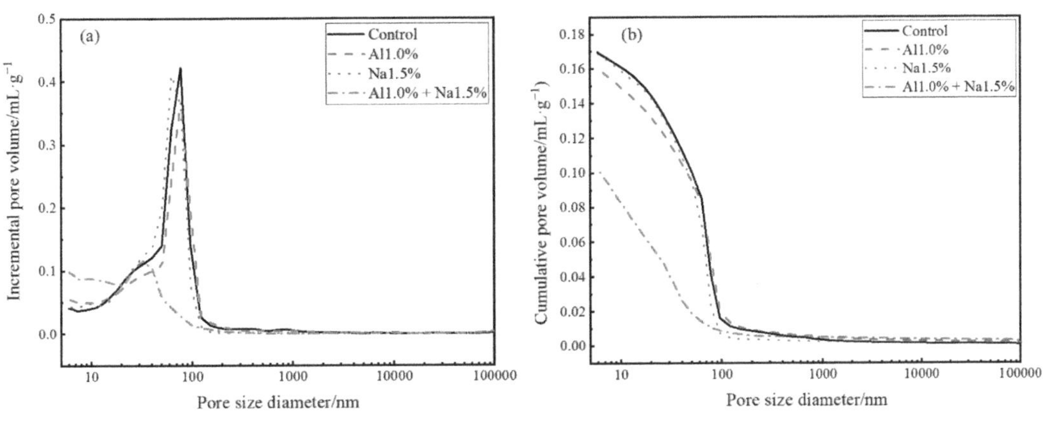

Figure 7. Effect of the different admixture Portland content on the pore structure of the hardened cement paste, (**a**) pore size distribution, (**b**) cumulative pore volume.

3.7. Discussion

Combined with previous studies and the macroscopic and microscopic test analysis of this experiment, we can infer that aluminum sulfate can promote the formation of AFt from

C_3A in cement paste, and the introduced Al^{3+} can react with Ca^{2+}, $(OH)^-$, SO_4^{2-} and water to form AFt. A large amount of AFt produced will crosslink to form the early skeleton of Portland cement paste. Compared to the control group, the considerable reduction of Ca^{2+} and $(OH)^-$ will rapidly promote the hydration of C_3S, then be weakened by the rapid decrease in C_3S, when again promote the hydration of C_3S and C_2S at after 48 h due to the compression of the C–S–H layer by Al^{3+} [3,20,21,43].

$$3CaO{\cdot}SiO_2 + nH_2O \rightarrow xCaO{\cdot}SiO_2{\cdot}yH_2O + (3 - x)Ca(OH)_2 + (n - 3 + x - y)H_2O \quad (6)$$

$$Al_2(SO_4)_3 + 3Ca(OH)_2 + 2H_2O \rightarrow 3CaSO_4{\cdot}2H_2O + 2Al(OH)_3{\downarrow} \quad (7)$$

$$3CaO{\cdot}Al_2O_3{\cdot}6H_2O + 3(CaSO_4{\cdot}2H_2O) + 6H_2O \rightarrow 3CaO{\cdot}Al_2O_3{\cdot}3CaSO_4{\cdot}32H_2O \quad (8)$$

$$2[Al(OH)_4]^- + 6Ca^{2+} + 4OH^- + 3SO_4^{2-} + 26H_2O \rightarrow 3CaO{\cdot}Al_2O_3{\cdot}3CaSO_4{\cdot}32H_2O \quad (9)$$

The addition of sodium sulfate to cement produces $CaSO_4{\cdot}2H_2O$, which promotes the hydration of C_3A to form AFt, the free Na^+ will combine with OH^- to form NaOH, which will increase the solubility of Ca^{2+} and promote the hydration of C_3S to form C–S–H [23,27,33]. However, soluble sodium salts in silicate cement paste will lose water and recrystallize in the later stage, which leads to swelling and cracking of Portland cement paste and affects its strength [28,51].

$$3CaO{\cdot}SiO_2 + nH_2O \rightarrow xCaO{\cdot}SiO_2{\cdot}yH_2O + (3 - x)Ca(OH)_2 + (n - 3+x - y)H_2O \quad (10)$$

$$Na_2SO_4 + 3Ca(OH_2) + 2H_2O \rightarrow 3CaSO_4{\cdot}2H_2O + 2NaOH \quad (11)$$

$$3CaO{\cdot}Al_2O_3{\cdot}6H_2O + 3(CaSO_4{\cdot}2H_2O) + 6H_2O \rightarrow 3CaO{\cdot}Al_2O_3{\cdot}3CaSO_4{\cdot}32H_2O \quad (12)$$

when aluminum sulfate and sodium sulfate are used synergistically, the introduced SO_4^{2-}, Al^{3+} and Na^+ work together to promote the generation of more AFt. The rapid reduction of Ca^{2+} and $(OH)^-$ in the Portland cement paste, the increased solubility of C_3S and C_2S by the generated NaOH, and the fast aggregation of C–S–H layers by the compression of Al^{3+}, all work together to promote the hydration of C_3S and C_2S in the Portland cement paste. According to Section 2.2.9, the Al1.0% + Na1.5% group had a smaller average pore size, and fewer harmful pores. The reason for this phenomenon is that more AFt cross-linking occurring early on makes the cement paste denser, and more hydration products such as C–S–H fill it. At the same time, Al^{3+} ions compress C–S–H, which reduces the porosity of the cement paste, reduces the average pore size, and makes the cement paste more compact, showing higher early compressive strength. Then, in the later stages there is still hydration of C_2S, which produces a continuous squeeze of C–S–H and makes the cement paste more compact.

4. Conclusions

In this work, different doses of aluminum sulfate and sodium sulfate, and the synergistic use of different doses of the DSAs as early strengthening agents were used to improve the performance of Portland cement paste. The setting and hardening properties of Portland cement paste were investigated with different admixtures and dosing ratios, the microstructure and hydration product composition of products were also analyzed separately. Finally, the mechanism of action when used singly and in combination was discussed. The results and discussion are summarized as follows.

(1) The early compressive strength of Portland cement paste at 3 d was significantly increased when using 1.0% of aluminum sulfate and 1.5% of sodium sulfate in combination, and the compressive strength at 28 d was also the best. The compressive strength at 3 d and 28 d were 40.4 MPa and 64.2 MPa, respectively, increasing 21.3% and 29.7% over the control group, 13.2% and 5.9% over the Al1.0% group, and 10.1% and 7.5% over Na1.5%, respectively.

(2) Compared to the control group, the use of aluminum sulfate promoted the hydration of C_3A, the reaction of Al^{3+}, Ca^{2+}, and $(OH)^-$ ions promoted the production of more AFt, while the reduction of Ca^{2+} and $(OH)^-$ ions rapidly promoted the hydration of C_3S. The addition of sodium sulfate promoted the generation of a certain amount of $CaSO_4 \cdot H_2O$, thus facilitating the hydration of C_3A to produce AFt, and the presence of Na^+ and $(OH)^-$ ions combined to promote the solubility of C_3S in the paste. When aluminum sulfate and sodium sulfate were used together, they synergistically promoted the hydration of C_3A to produce AFt, facilitating the production of more AFt and the hydration of C_3S and C_2S to produce more C–S–H.

(3) The synergistic use of 1.0% of aluminum sulfate and 1.5% of sodium sulfate resulted in the generation of more AFt in the Portland cement paste, which was cross-linked together to form the early skeleton of Portland cement paste base, and the products generated such as C–S–H, $CaCO_3$ and $CaSO_4 \cdot H_2O$ were filled in. Compared to the control group, the Al1.0% group, and the Na1.5% group, the Al1.0% + Na1.5% group resulted in a denser structure with reduced porosity, fewer harmful pores and more harmless pores, which allowed the Portland cement paste to exhibit a higher compressive strength.

Author Contributions: Conceptualization, L.S.; data curation, L.S.; formal analysis, Y.W.; funding acquisition, Y.W. and S.L. (Songhui Liu); investigation, L.S. and S.L (Shuaijie Li); methodology, Y.W., S.L. (Songhui Liu), S.L. (Shuqiong Luo) and X.G.; project administration, Y.W.; resources, Y.W., S.L. (Songhui Liu), X.G. and S.L. (Shuqiong Luo); supervision, Y.W. and S.L. (Songhui Liu); validation, L.S. and S.L. (Shuaijie Li); writing—original draft, L.S.; writing—review and editing, Y.W., L.S., S.L. (Songhui Liu), X.G. and S.L. (Shuqiong Luo). All authors have read and agreed to the published version of the manuscript.

Funding: This work was supported by the National Natural Science Foundation of China (52108208), the fellowship of China Postdoctoral Science Foundation (2020M682290), the science and technology project of Henan Province (211110231400, 212102310559), the National Key R&D Program Intergovernmental International Science and Technology Innovation Cooperation Project (2018YFE0107300), the Opening Project of State Key Laboratory of Green Building Materials (2021GBM06), the Henan Outstanding Foreign Scientists' Workroom (GZS2021003), the doctor foundation of Henan Polytechnic University (B2020-11), and the Fundamental Research Funds for the Universities of Henan Province (NSFRF220302).

Institutional Review Board Statement: Not applicable.

Informed Consent Statement: Not applicable.

Data Availability Statement: Not applicable.

Conflicts of Interest: The authors declare no conflict of interest.

References

1. Aggoun, S.; Cheikh-Zouaoui, M.; Chikh, N.; Duval, R. Effect of some admixtures on the setting time and strength evolution of cement pastes at early ages. *Constr. Build. Mater.* **2008**, *22*, 106–110. [CrossRef]
2. Wu, M.; Zhang, Y.; Jia, Y.; She, W.; Liu, G.; Yang, Z.; Zhang, Y.; Zhang, W.; Sun, W. Effects of sodium sulfate on the hydration and properties of lime-based low carbon cementitious materials. *J. Clean. Prod.* **2019**, *220*, 677–687. [CrossRef]
3. Liu, S.; Shen, Y.; Wang, Y.; He, H.; Luo, S.; Huang, C. Synergistic use of sodium bicarbonate and aluminum sulfate to enhance the hydration and hardening properties of Portland cement paste. *Constr. Build. Mater.* **2021**, *299*, 124248. [CrossRef]
4. Heren, Z.; Ölmez, H. The influence of ethanolamines on the hydration and mechanical properties of Portland cement. *Cem. Concr. Res.* **1996**, *26*, 701–705. [CrossRef]
5. Bao, J.; Ren, Q.; Sun, L.; Ding, Y.; Oh, W.-C. Preparation of an Early Strengthening Agent for Concrete under Low-Temperature Conditions and Evaluation of Its Reaction Mechanism. *Korean J. Mater. Res.* **2021**, *31*, 195–208. [CrossRef]
6. Lee, T.; Lee, J.; Choi, H.; Lee, D. The Effects of Fineness and TEA-Based Chemical Admixture on Early Strength Development of Concrete in Construction Site Applications. *Materials* **2020**, *13*, 2027. [CrossRef]
7. Izotov, V.S.; Ibragimov, R.A. Hydration Products of Portland Cement Modified with a Complex Admixture. *Inorg. Mater.* **2015**, *2*, 87–190. [CrossRef]

8. Izotov, V.S.; Ibragimov, R.A. Influence of additives-hardening accelerators on the properties of heavy concrete. *Stroit. Mater.* **2010**, *3*, 35–37.
9. Fakhri, M.; Yousefian, F.; Amoosoltani, E.; Aliha, M.R.M.; Berto, F. Combined effects of recycled crumb rubber and silica fume on mechanical properties and mode I fracture toughness of self-compacting concrete. *Fatigue Fract. Eng. Mater. Struct.* **2021**, *44*, 2659–2673. [CrossRef]
10. Fakhri, M.; Amoosoltani, E. Crack behavior analysis of roller compacted concrete mixtures containing reclaimed asphalt pavement and crumb rubber. *Eng. Fract. Mech.* **2017**, *180*, 43–59. [CrossRef]
11. Fakhri, M.; Amoosoltani, E. The effect of Reclaimed Asphalt Pavement and crumb rubber on mechanical properties of Roller Compacted Concrete Pavement. *Constr. Build. Mater.* **2017**, *137*, 470–484. [CrossRef]
12. Fakhri, M.; Saberi, K.F. The effect of waste rubber particles and silica fume on the mechanical properties of Roller Compacted Concrete Pavement. *J. Clean. Prod.* **2016**, *129*, 521–530. [CrossRef]
13. Jiang, M.; Lv, X. Progress in Research and Application of early strength of concrete. *Bull. Chin. Silic. Soc.* **2014**, *33*, 2527–2533.
14. Chen, S.; Du, Z.; Zhang, Z.; Zhang, H.; Xia, Z.; Feng, F. Effects of chloride on the early mechanical properties and microstructure of gangue-cemented paste backfill. *Constr. Build. Mater.* **2020**, *235*, 117504. [CrossRef]
15. Wang, J.; Qian, C.; Qu, J.; Guo, J. Effect of lithium salt and nano nucleating agent on early hydration of cement based materials. *Constr. Build. Mater.* **2018**, *174*, 24–29. [CrossRef]
16. Lu, C.; Yuan, S.; Cheng, P.; Liu, R. Mechanical properties of corroded steel bars in pre-cracked concrete suffering from chloride attack. *Constr. Build. Mater.* **2016**, *123*, 649–660. [CrossRef]
17. Ren, G.; Tian, Z.; Wu, J.; Gao, X. Effects of combined accelerating admixtures on mechanical strength and microstructure of cement mortar. *Constr. Build. Mater.* **2021**, *304*, 124642. [CrossRef]
18. Paglia, C.; Wombacher, F.; Böhni, H. The influence of alkali-free and alkaline shotcrete accelerators within cement systems: I. Characterization of the setting behavior. *Cem. Concr. Res.* **2001**, *31*, 913–918. [CrossRef]
19. Prudêncio, L.R. Accelerating admixtures for shotcrete. *Cem. Concr. Compos.* **1998**, *20*, 213–219. [CrossRef]
20. Yang, R.; He, T.; Guan, M.; Guo, X.; Xu, Y.; Xu, R.; Da, Y. Preparation and accelerating mechanism of aluminum sulfate-based alkali-free accelerating additive for sprayed concrete. *Constr. Build. Mater.* **2020**, *234*, 117334. [CrossRef]
21. Liu, X.; Ma, B.; Tan, H.; Gu, B.; Zhang, T.; Chen, P.; Li, H.; Mei, J. Effect of aluminum sulfate on the hydration of Portland cement, tricalcium silicate and tricalcium aluminate. *Constr. Build. Mater.* **2020**, *232*, 117179. [CrossRef]
22. Chen, C.; Sun, Z. Influence of Aluminum Sulfate on Hydration and Properties of Cement Pastes. *J. Adv. Concr. Technol.* **2018**, *16*, 522–530. [CrossRef]
23. Li, H.; Xue, Z.; Liang, G.; Wu, K.; Dong, B.; Wang, W. Effect of C–S–H s-PCE and sodium sulfate on the hydration kinetics and mechanical properties of cement paste. *Constr. Build. Mater.* **2021**, *266*, 121096. [CrossRef]
24. Zou, F.; Hu, C.; Wang, F.; Ruan, Y.; Hu, S. Enhancement of early-age strength of the high content fly ash blended cement paste by sodium sulfate and C–S–H seeds towards a greener binder. *J. Clean. Prod.* **2020**, *244*, 118566. [CrossRef]
25. Fu, J.; Jones, A.M.; Bligh, M.W.; Holt, C.; Keyte, L.M.; Moghaddam, F.; Foster, S.J.; Waite, T.D. Mechanisms of enhancement in early hydration by sodium sulfate in a slag-cement blend—Insights from pore solution chemistry. *Cem. Concr. Res.* **2020**, *135*, 106110. [CrossRef]
26. Zhao, Y.; Bai, H. Effect of early-strength admixture for concrete performance. *LiaoNing Build. Mater.* **2009**, *7*, 19–21.
27. Tan, H.; Deng, X.; He, X.; Zhang, J.; Zhang, X.; Su, Y.; Yang, J. Compressive strength and hydration process of wet-grinded granulated blast-furnace slag activated by sodium sulfate and sodium carbonate. *Cem. Concr. Compos.* **2019**, *97*, 387–398. [CrossRef]
28. Campos, A.; López, C.M.; Blanco, A.; Aguado, A. Effects of an internal sulfate attack and an alkali-aggregate reaction in a concrete dam. *Constr. Build. Mater.* **2018**, *166*, 668–683. [CrossRef]
29. *GB175-2007*; Common Portland Cement. Standardization Administration of China: Beijing, China, 2017.
30. *GB/T 1346-2011*; Test Methods for Water Requirement of Normal Consistency, Setting Time and Soundness of the Portland Cement. Standardization Administration of China: Beijing, China, 2011.
31. *GB/T 50081-2019*; Standard for Test Method of Concrete Physical and Mechanical Properties. Standardization Administration of China: Beijing, China, 2019.
32. Ma, B.; Zhang, T.; Tan, H.; Liu, X.; Mei, J.; Qi, H.; Jiang, W.; Zou, F. Effect of triisopropanolamine on compressive strength and hydration of cement-fly ash paste. *Constr. Build. Mater.* **2018**, *179*, 89–99. [CrossRef]
33. Mei, J.; Tan, H.; Li, H.; Ma, B.; Liu, X.; Jiang, W.; Zhang, T.; Li, X. Effect of sodium sulfate and nano-SiO_2 on hydration and microstructure of cementitious materials containing high volume fly ash under steam curing. *Constr. Build. Mater.* **2018**, *163*, 812–825. [CrossRef]
34. Dai, J.; Wang, Q.; Lou, X.; Bao, X.; Zhang, B.; Wang, J.; Zhang, X. Solution calorimetry to assess effects of water-cement ratio and low temperature on hydration heat of cement. *Constr. Build. Mater.* **2021**, *269*, 121222. [CrossRef]
35. Zhang, J.; Li, G.; Ye, W.; Chang, Y.; Liu, Q.; Song, Z. Effects of ordinary Portland cement on the early properties and hydration of calcium sulfoaluminate cement. *Constr. Build. Mater.* **2018**, *186*, 1144–1153. [CrossRef]
36. Luo, S.Q.; Zhao, M.H.; Jiang, Z.Z.; Liu, S.H.; Yang, L.; Mao, Y.X.; Pan, C.G. Microwave preparation and carbonation properties of low-carbon cement. *Constr. Build. Mater.* **2021**, *320*, 126239. [CrossRef]

37. Guan, X.; Liu, S.; Feng, C.; Qiu, M. The hardening behavior of γ-C2S binder using accelerated carbonation. *Constr. Build. Mater.* **2016**, *114*, 204–207. [CrossRef]
38. Jain, J.; Neithalath, N. Analysis of calcium leaching behavior of plain and modified cement pastes in pure water. *Cem. Concr. Compos.* **2009**, *31*, 176–185. [CrossRef]
39. Criado, M.; Jiménez, A.F.; Palomo, A. Effect of sodium sulfate on the alkali activation of fly ash. *Cem. Concr. Compos.* **2010**, *32*, 589–594. [CrossRef]
40. Andrade Neto, J.S.; de Matos, P.R.; De la Torre, A.G.; Campos, C.E.M.; Gleize, P.J.P.; Monteiro, P.J.M.; Kirchheim, A.P. The role of sodium and sulfate sources on the rheology and hydration of C3A polymorphs. *Cem. Concr. Res.* **2022**, *151*, 106639. [CrossRef]
41. Joseph, S.; Skibsted, J.; Cizer, Ö. A quantitative study of the C3A hydration. *Cem. Concr. Res.* **2019**, *115*, 145–159. [CrossRef]
42. Quennoz, A.; Scrivener, K.L. Hydration of C3A–gypsum systems. *Cem. Concr. Res.* **2012**, *42*, 1032–1041. [CrossRef]
43. Brykov, A.S.; Vasil Ev, A.S.; Mokeev, M.V. Hydration of portland cement in the presence of aluminum-containing setting accelerators. *Russ. J. Appl. Chem.* **2013**, *86*, 793–801. [CrossRef]
44. He, H.; Wang, Y.; Wang, J. Compactness and hardened properties of machine-made sand mortar with aggregate micro fines. *Constr. Build. Mater.* **2020**, *250*, 118828. [CrossRef]
45. Liu, S.; Zhang, H.; Wang, Y.; Gou, M. Carbon-dioxide-activated bonding material with low water demand. *Adv. Cem. Res.* **2021**, *33*, 193–196. [CrossRef]
46. Liu, S.; Shen, P.; Xuan, D.; Li, L.; Sojobi, A.; Zhan, B.; Poon, C.S. A comparison of liquid-solid and gas-solid accelerated carbonation for enhancement of recycled concrete aggregate. *Cem. Concr. Compos.* **2021**, *118*, 103988. [CrossRef]
47. Rashad, A.M.; Bai, Y.; Basheer, P.A.M.; Milestone, N.B.; Collier, N.C. Hydration and properties of sodium sulfate activated slag. *Cem. Concr. Compos.* **2013**, *37*, 20–29. [CrossRef]
48. Cai, J. The correct application of chemical equilibrium Le chatelier's principle to judge the direction of movement. *Fujian Basic Educ. Res.* **2013**, *11*, 40–41.
49. Lin, Y. Correctly Understanding and Using Le chatelier's principle. *Chem. Educ.* **2012**, *33*, 3.
50. Wang, L.; Yang, H.; Zhou, S.; Chen, E.; Tang, S. Hydration, mechanical property and C–S–H structure of early-strength low-heat cement-based materials. *Mater. Lett.* **2018**, *217*, 151–154. [CrossRef]
51. Comi, C.; Perego, U. Anisotropic Damage Model for Concrete Affected by Alkali-Aggregate Reaction. *Int. J. Damage Mech.* **2011**, *4*, 598–617. [CrossRef]
52. Owsiak, Z.; Zapała-Sławeta, J.; Czapik, P. Diagnosis of concrete structures distress due to alkali-aggregate reaction. *Bull. Pol. Acad. Sci. Tech. Sci.* **2015**, *63*, 23–29. [CrossRef]
53. Zhang, B.; Tan, H.; Ma, B.; Chen, F.; Lv, Z.; Li, X. Preparation and application of fine-grinded cement in cement-based material. *Constr. Build. Mater.* **2017**, *157*, 34–41. [CrossRef]
54. Pang, B.; Zhou, Z.; Xu, H. Utilization of carbonated and granulated steel slag aggregate in concrete. *Constr. Build. Mater.* **2015**, *84*, 454–467. [CrossRef]
55. Xue, J.; Liu, S.; Ma, X.; Teng, Y.; Guan, X. Effect of different gypsum dosage on the chloride binding properties of C4AF hydrated paste. *Constr. Build. Mater.* **2022**, *315*, 125562. [CrossRef]

coatings

Article

Evaluation of Half-Cell Potential Measurements for Reinforced Concrete Corrosion

Yousef Almashakbeh [1,*], Eman Saleh [2] and Nabil M. Al-Akhras [3]

[1] Department of Allied Engineering Sciences, Faculty of Engineering, The Hashemite University, P.O. Box 330127, Zarqa 13133, Jordan
[2] Civil Engineering Department, Faculty of Engineering, The Hashemite University, P.O. Box 330127, Zarqa 13133, Jordan; emanf_yo@hu.edu.jo
[3] Civil Engineering Department, Jordan University of Science and Technology, P.O.Box 3030, Irbid 22110, Jordan; alakhras@just.edu.jo
* Correspondence: yousefalmashakbeh@hu.edu.jo

Abstract: The evaluation of half-cell potential measurements in reinforced concrete (RC) members can be a key issue for civil engineers. The primary reason for this is that the interpretation of half-cell potential measurements based on the available standards provides information related only to the possibility of corrosion in concrete, but it does not provide a clear perception of the influence of corrosion on the capacity of the RC members. The objective of this study is two-fold: (1) to explore the influence of corrosion level on the flexural capacity of RC members; and (2) to provide engineers with a better understanding of the correlation between half-cell potential measurements and flexural capacity of RC members. To establish this, twelve RC beams were cast and then exposed to accelerated corrosion utilizing an impressed current. After that, half-cell potential tests were performed on the entire surface of the beams. Next, a four-point loading test was performed on the beams to determine their flexural behavior. The analysis of measurements showed that there is a high positive correlation between the half-cell potential measurements and the flexural capacity of the tested beams which demonstrates the potential of half-cell measurements to predict the capacity degradation level of the RC beams due to corrosion.

Keywords: corrosion; half-cell potential; ductility; flexural capacity

1. Introduction

Degradation of structures due to corrosion of embedded steel is considered one of the major durability problems faced by the construction industry [1–5]. The rate of steel corrosion is increased by the ingression of either chloride ion (Cl^-) or CO_2 into concrete, reducing the alkaline hydrated cement products that passivate (i.e., shield) the embedded steel. Corrosion causes a reduction in the cross-section of embedded steel, thereby reducing its flexural and tensile strength capacities [6,7]. This could cause the development of many cracks on the concrete that, in turn, reduce the load capacity of the structure.

The *NACE* (2013) impact study reported that the estimated cumulative cost of corrosion was approximately 4% of the global gross domestic product. Further, it reported that the utilization of corrosion prevention practices may yield a reduction in the cost by 15–35% [8]. The deteriorated structures due to corrosion may require a higher cost for rehabilitation than the cost of initial construction [9–11]. Moreover, the corrosion of embedded steel could increase the risk of failure due to the presence of external forces such as earthquakes. Therefore, an early detection and assessment of reinforced concrete (RC) corrosion are necessary to accelerate corrective measures [12,13]. This assessment should provide information regarding the remaining service life, level of damage, and required repairs [14,15].

The half-cell potential method is a commonly applied technique for detecting corrosion in RC members [16,17]. This technique is performed and interpreted based on ASTM C876 standards [18]. It measures the value of potential on the surface of concrete members where the positive pole of the high-impedance volumeter is electrically attached to the reinforcing steel and the other pole is attached to a reference electrode. The measured potential is then translated into the risk of corrosion of embedded steel inside the RC members [1,19]. Table 1 shows the criteria followed to interpret the potential values for corrosion detection. For example, when the potential value is less than −200 mV, the reinforced steel is categorized to have a low risk of corrosion (i.e., less than 10% risk level).

Table 1. Corrosion activity according to half-cell values (ASTM C876).

Half-Cell Potential Values	Probability of Corrosion
less negative than −200 mV	10% probability of corrosion
between −200 mV and −350 mV	Approximately 50%
more negative than −350 mV	90% probability of corrosion

Few studies have considered the half-cell potential test as an indicator of corrosion in embedded reinforcing bars. Jung et al. [11] and Gonzalez et al. [20] performed an experimental study to evaluate the structural performance of corrosion-damaged RC beams. In their study, it was concluded that corrosion adversely affects the ultimate load and energy dissipation capacity of RC members and they suggested further experimentation for evaluating the performance of structures with corrosion-damaged members. Naish et al. [21] and Pradhan and Bhattacharjee [22] tried to link the potential and resistivity measurements to the corrosion of RC members. According to these studies, a strong correlation between the half-cell potential measurements and the corrosion level was observed; however, they did not link the half-cell potential measurements to the performance degradation level of corrosion-damaged members. Adriman et al. [19] showed, using a case study, that a half-cell potential test augmented with computational approaches is a promising method to improve the diagnosis of corrosion in RC members. Thus, Adriman et al. [19] demonstrated that half-cell potential measurements can be used beyond the qualitative corrosion risk assessment established by the ASTM C876 [18] (see Table 1).

In this study, the objective is to quantify the impact of corrosion on the flexural behavior of RC beams and link it to half-cell potential measurements. This link will help to expand the usability of half-cell potential measurements to quantitatively predict the degree of degradation of corrosion-damaged members. In addition, the effectiveness of half-cell potential measurements as corrosion indicators will be investigated through this established link.

2. Experimentation

Figure 1 illustrates the process followed to prepare the control beams and corroded beams used to establish the link between the flexural behavior of corrosion-damaged beams and half-cell potential measurements. As shown in Figure 1, first, the concrete mixture and reinforced cage were designed, based on the ACI-211 guidelines [23] and ACI 318 standards [24], respectively. After that, the mixture was prepared, tested for workability and strength (i.e., slump and cube test, respectively), and cast into forms. To induce corrosion, an electrically impressed current was used. After prespecified durations, half-cell potential tests were performed on the entire surface of the beams. Then, a four-point loading test was conducted to estimate the impact of corrosion on the flexural behavior of RC beams. Finally, the link was established by a comparison between the strength-deformation properties of corroded beams, obtained from the four-point loading test, and the half-cell potential measurements. The following is a detailed description of each step illustrated in Figure 1.

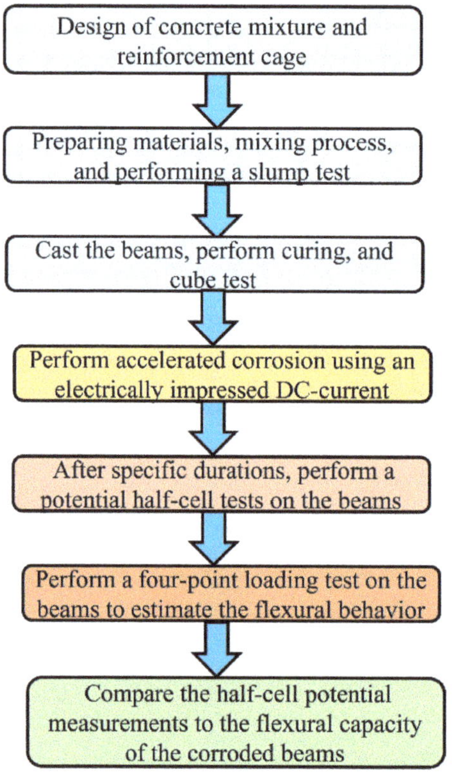

Figure 1. Schematic flow chart used for the evaluation of half-cell potential measurement.

2.1. Material

The concrete mix was designed following the ACI-211 guidelines [23], with a water-to-cement ratio of 0.47. The RC beams were prepared using ordinary Portland cement (Type I) with coarse limestone aggregates of 12.5 mm, in addition to a mixture of silica sands and crushed fine limestone (20% silica and 80% fine aggregate). The ASTM standard technique C136 was used to determine the fineness modulus of fine limestone. Following the ASTM C127 method, the saturated surface dry (SSD) bulk specific gravity (BSG) and absorption of coarse limestone aggregate were 2.56 and 1.34%, respectively. Moreover, according to ASTM C128, the bulk specific gravity (SSD) and absorption of fine limestone aggregate were 2.52 and 3.41%, respectively. For silica sand, the corresponding percentages were 2.59 and 0.73%. Further, as per ASTM C29, the unit weight for coarse aggregate was 1362 kg/m^3. The beam reinforcements were constructed using two distinct grades of steel reinforcement: G40 and G60. The mechanical characteristics of the steel reinforcements are listed in Table 2.

Table 2. Mechanical properties of steel reinforcement.

Size	Grade	Yield Strength, Fy (MPa)	Ultimate Strength, F_u (MPa)	Elongation %
6 mm	G40	290	353	14
10 mm	G60	441.8	657	12.9
12 mm	G60	435.5	631	14.7
14 mm	G60	463	640	16

2.2. *Test Specimens*

Twelve reinforced concrete beams with dimensions 150 × 100 × 1000 mm were tested. All rebars were 12 mm in diameter and embedded at 35 cm cover thickness from the surface. Stirrups with a spacing of 50 mm were provided along the length of the beam. Internally dimensioned wooden molds were constructed and used to cast the test specimens. Before concrete casting, the steel reinforcement was assembled and placed within the molds in the shape of a cage. Three beams and three cylinders (100 × 200 mm) were cast in each batch to be tested for compressive strength, which was determined to be around 41.6 MPa. A tilting drum mixer with a batch volume of approximately a quarter m^3 was used. The slump was tested (per ASTM-C143) and determined to be about 50 mm. Three layers of fresh concrete were poured into the wooden molds. Each layer was vibrated on a compacting table. After 24 h, the beam specimens were demolded and cured for 28 days with wet burlap. Figure 2 shows the above steps.

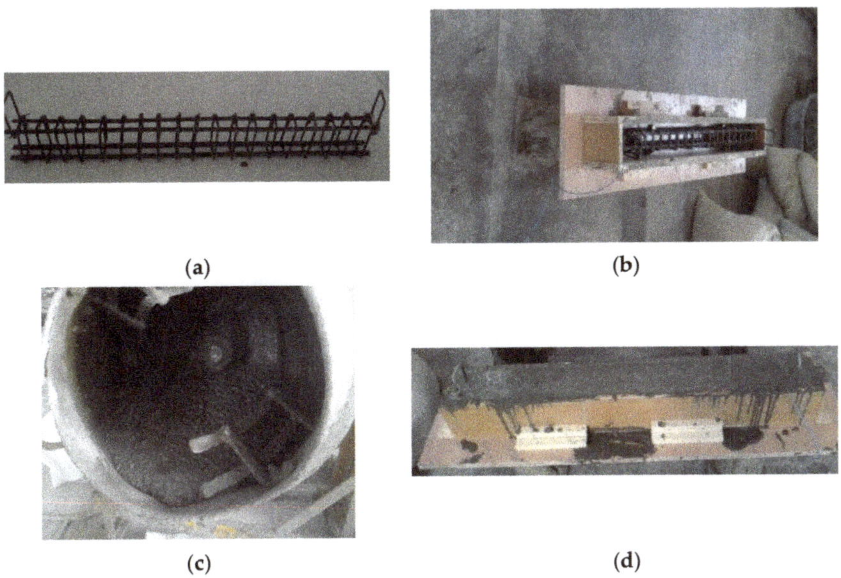

Figure 2. Steps for preparation of test specimens. (**a**) A typical steel reinforcement. (**b**) Steel reinforcement cage inside the wooden mold. (**c**) Mixing of concrete ingredients. (**d**) Casted specimens.

2.3. *Corrosion Process*

The electrically impressed DC was used to accelerate the corrosion process on the RC beams. The accelerated corrosion process was achieved by applying DC (direct current) to both the longitudinal steel bars and stirrups, which serve as anodes within each beam through metallic wires put before casting. The cathode was constructed using a steel plate with a dimension of (800 × 300 × 10 mm) that was linked to the cathode wire. The specimens were subjected to a 200 $\mu A/cm^2$ current density. To create an aggressive environment, samples were put in a tank containing a solution of NaCl at a concentration of 2%, as illustrated in Figure 3.

(a) (b)

Figure 3. The accelerated corrosion process. (**a**) Three days of accelerated corrosion. (**b**) Eight days of accelerated corrosion.

2.4. Half-Cell Potential Test

Nondestructive tests were used to analyze the physical qualities of concrete, such as discontinuities and changes in the composition of concrete and corrosion level, without causing any damage to the tested specimens. Here, the half-cell potential test was performed on the cast RC beams. Each beam was split into ten sections. The half-cell potential equipment was used to collect three readings at each test location per the ASTM C876 [18] requirements. The measurement was performed using a silver/silver chloride reference electrode. Before the test, the beam was moistened with water to enhance the moisture content of the concrete. Then, a saturated sponge with alcohol was put on the concrete surface, followed by the reference electrode. The other end of the voltmeter was then linked to the steel rebar (see Figure 4).

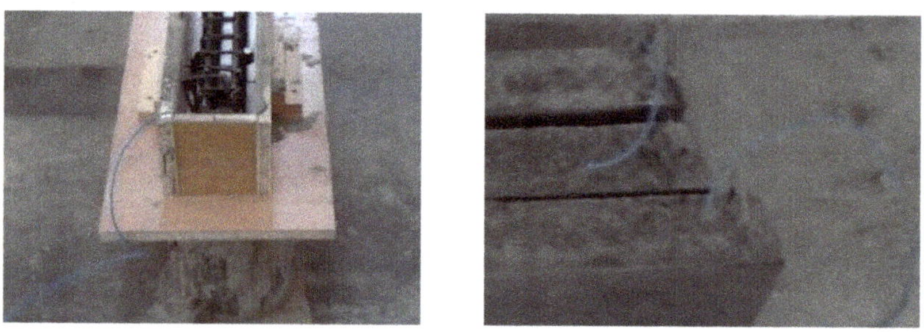

(a) (b)

Figure 4. Connected wire with rebar. (**a**) Before casting. (**b**) After casting.

2.5. Flexural Behavior

Flexural testing was performed on the beams under four-point loading across a span of 0.9 m in a simple support arrangement. Between the two loading points, the spacing was 300 mm. The beam was supported by a special sort of support made of hardened steel to guarantee that no deformation occurred that might impact the test findings. All beams were loaded with two-point loads utilizing 400 kN hydraulic testing equipment at a loading rate of 0.3 kN/s. Supports were designed to function as a hinge on one end and a roller on the other. The vertical deflection at the mid-span was determined using a linear variable displacement transducer (LVDT). The load-deflection curves were constructed using the data obtained from the load cell and LVDT. The test setup is illustrated in Figure 5.

Figure 5. The test set-up used.

3. Experimental Results and Discussion

This section presents and discusses the experimental results of the performed experiments. An evaluation of half-cell potential measurements was performed by comparison with the flexural capacity of the corrosion-damaged RC beams.

In the tests, all the beams (i.e., uncorroded and corroded) initially developed flexural cracks in the high moment zone region (see Figure 6), which subsequently expanded into the compression zone and propagated across the shear span as the load increased. Before collapsing, the cracks expanded deep into the compression zone. As shown in Figure 6, the crack patterns for both the control beams and corroded beams were very similar and exhibited a ductile failure (i.e., flexural failure).

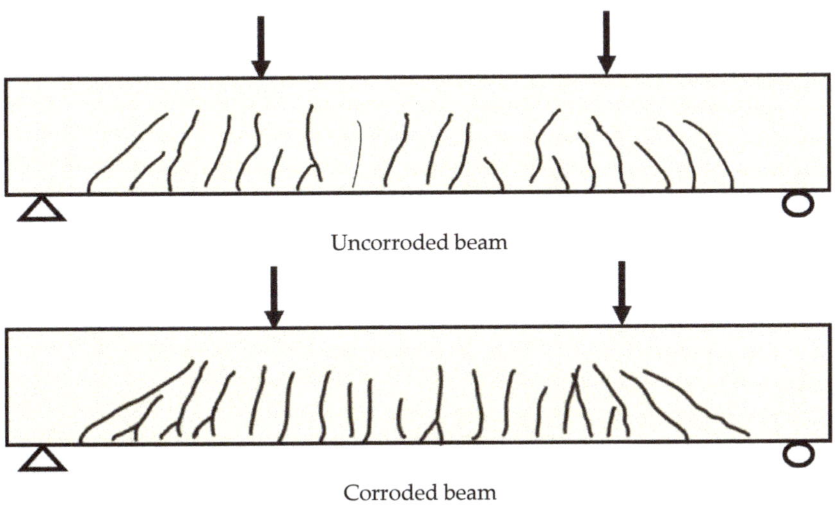

Figure 6. Crack patterns of the uncorroded and corroded beam.

The load and vertical deflection of the beams were recorded and plotted for each specimen during the test. These curves were then investigated based on their initial stiffness, maximum displacement, maximum load, and toughness.

3.1. Flexural Behavior

The mechanical response of the beams was determined using load-deflection curves and their related properties, including maximum displacement, maximum load, initial stiffness, and toughness. The load-deflection curves and the average features of the curves of six of the twelve tested corroded beams and their corresponding uncorroded RC beam (i.e., control beams) are illustrated in Figure 7 and Table 3, respectively. As shown in Figure 7, the curves exhibited linear behavior during the first phase of loading before showing a nonlinear behavior. All the corroded beams were found to have a lower ultimate capacity than the control beams. The average load capacity of all corroded beams was computed to be 84% of the control beams. In addition, the average ductility of corroded beams was estimated to be 114.4% of the average ductility of the corresponding control beams. All corroded RC beams had a greater stiffness than uncorroded beams; the corroded beams retained 111% of their original stiffness. These observations are similar to the results observed in [11,20] where it was concluded that corrosion adversely affects the flexural capacity of RC beams.

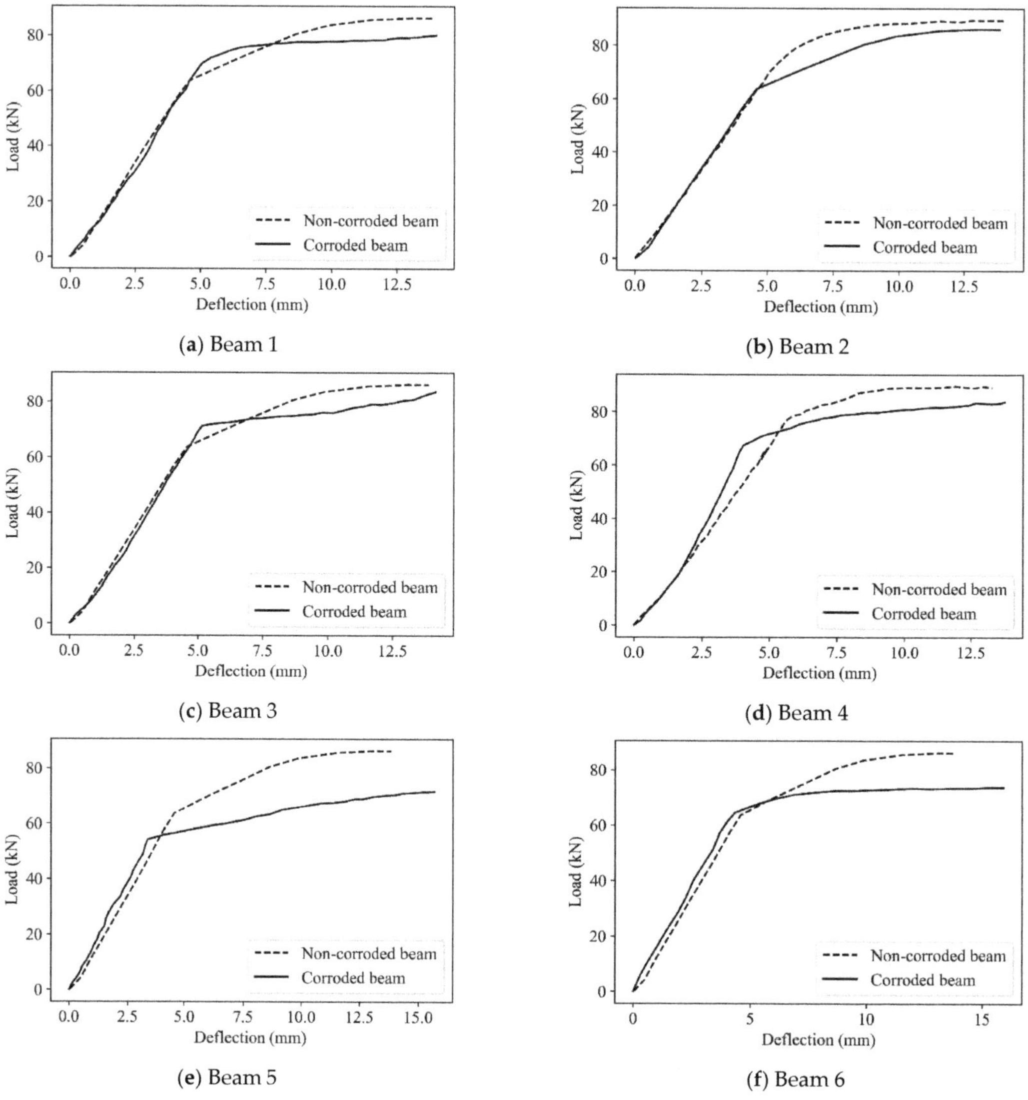

(a) Beam 1; (b) Beam 2; (c) Beam 3; (d) Beam 4; (e) Beam 5; (f) Beam 6

Figure 7. Load–deflection curves for the RC beams.

Table 3. The average mechanical characteristics for control and corroded beams.

Case	Beam Designation	Ultimate Load (kN)	Max Deflection (mm)	Toughness (J)	Stiffness (MN/m)
1	Control beam	86.2	13.9	877.5	13.75
2	Corroded beams	72.7 (84%) *	15.80 (114%) *	928.0 (106%) *	15.21 (111%) *

* Residual properties.

3.2. Analysis of Half-Cell Potential Measurements

Figure 8 illustrates the half-cell potential values for one of the tested RC beams before and after corrosion exposure. Before corrosion, all beams had values less than −200 mV, indicating a low possibility of corrosion (see Table 1). However, after corrosion, the half-cell potential measurements increased drastically, exceeding −350 mV, indicating the presence of considerable corrosion. The other tested RC beams showed similar behavior to the one in Figure 8.

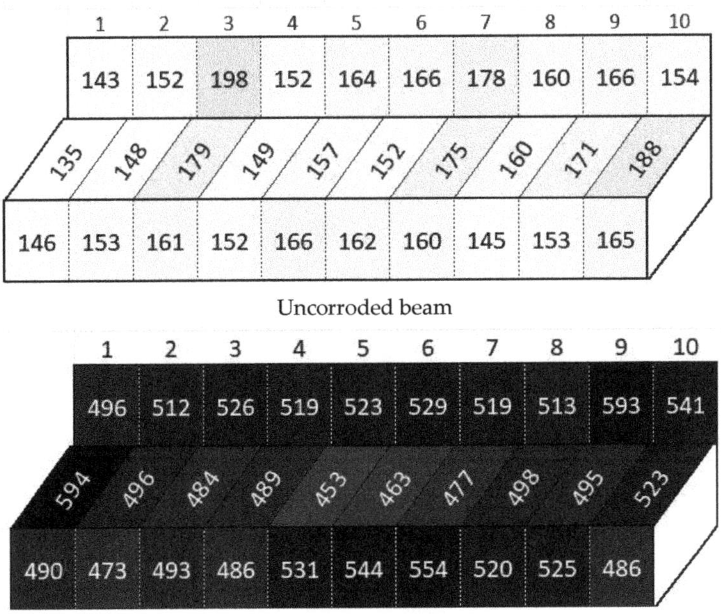

Figure 8. Half-cell potential measurements of the corroded beams after 8 days of corrosion and the corresponding uncorroded one.

3.3. Correlation between Half-Cell Potential Measurements and Four-Point Loading Test

Performing a correlation study between the half-cell potential measurements and the four-point loading test resulted in a negative correlation between the half-cell potential measurements and the ultimate load capacity of corroded RC beams and a positive correlation between the half-cell potential measurements and the ductility of half-cell potential measurements. A Pearson's correlation coefficient of −0.93 indicated a high correlation between the half-cell potential measurements and flexural capacity of RC beams, while a Pearson's correlation coefficient of 0.74 between the half-cell potential measurements and ductility of the beams implied a high-to-moderate correlation between the results.

4. Conclusions and Recommendations for Future Research

In this study, the effect of corrosion of embedded steel on the flexural behavior of RC beams and the level of correlation between half-cell potential measurements and the flexural behavior of RC beams were investigated. The following results can be drawn from the experimental investigation:

- The objective of this study was to provide a better understanding of the strength-deformation behavior of corroded members. Based on the test results, it was observed that the failure patterns of the corroded beams were generally similar to the control beams. However, it was observed that the ultimate capacity of the RC beams decreased, while the stiffness and ductility increased with the duration of accelerated corrosion.
- The ductility increased with increasing half-cell potential, while the flexural capacity decreased with increasing half-cell potential. The increase in ductility was 114.4% of residual deflection and the decrease in flexural capacity was 16% lower than that of control beams at the maximum half-cell potential.
- We observed a relatively strong correlation between the average potential difference and the degradation in the flexural capacity of the beams (Pearson's correlation = 0.93. Thus, the potential difference can be used as an indicator of the degradation level in RC members.
- To propose a practical procedure for evaluating the flexural capacity of corroded members using half-cell potential measurements, further experimental research on corroded members, together with analytical analysis such as finite element analysis, should be performed. This practical procedure may aspire to derive a degradation factor formula that uses half-cell potential measurements to predict the capacity of corrosion-damaged members.

Author Contributions: Conceptualization, Y.A.; data curation and elaboration, E.S.; experimental tests, Y.A.; validation and writing, Y.A., E.S. and N.M.A.-A. All authors have read and agreed to the published version of the manuscript.

Funding: This research was funded by the Deanship of Scientific Research at Jordan University of Science and Technology (Project 100/2016).

Institutional Review Board Statement: Not applicable.

Informed Consent Statement: Not applicable.

Data Availability Statement: The data is available upon reasonable request from the authors.

Conflicts of Interest: The authors declare no conflict of interest.

References

1. Yodsudjai, W.; Pattarakittam, T. Factors influencing half-cell potential measurement and its relationship with corrosion level. *Measurement* **2017**, *104*, 159–168. [CrossRef]
2. Qiao, G.; Hong, Y.; Ou, J. Quantitative monitoring of pitting corrosion based on 3-D cellular automata and real-time ENA for RC structures. *Measurement* **2014**, *53*, 270–276. [CrossRef]
3. Li, C.; Chen, Q.; Wang, R.; Wu, M.; Jiang, Z. Corrosion assessment of reinforced concrete structures exposed to chloride environments in underground tunnels: Theoretical insights and practical data interpretations. *Cem. Concr. Compos.* **2020**, *112*, 103652. [CrossRef]
4. Abouhussien, A.A.; Hassan, A.A.A. Acoustic emission monitoring of corrosion damage propagation in large-scale reinforced concrete beams. *J. Perform. Constr. Facil.* **2018**, *32*, 4017133. [CrossRef]
5. Deng, H.; Zhu, P.; Hu, C.; He, T. Study on Dynamic Lubrication Characteristics of the External Return Spherical Bearing Pair under Full Working Conditions. *Machines* **2022**, *10*, 107. [CrossRef]
6. Zhou, Y.; Gencturk, B.; Willam, K.; Attar, A. Carbonation-induced and chloride-induced corrosion in reinforced concrete structures. *J. Mater. Civ. Eng.* **2015**, *27*, 4014245. [CrossRef]
7. Meng, D.; Lin, S.; Azari, H. Nondestructive corrosion evaluation of reinforced concrete bridge decks with overlays: An experimental study. *J. Test. Eval.* **2019**, *48*, 516–537. [CrossRef]
8. Koch, G.; Varney, J.; Thompson, N.; Moghissi, O.; Gould, M.; Payer, J. International measures of prevention, application, and economics of corrosion technologies study. *NACE Int.* **2016**, *216*, 2–3.

9. Qiao, G.; Hong, Y.; Ou, J. Corrosion monitoring of the RC structures in time domain: Part I. Response analysis of the electrochemical transfer function based on complex function approximation. *Measurement* **2015**, *67*, 78–83. [CrossRef]
10. Zhang, J.; Liu, C.; Sun, M.; Li, Z. An innovative corrosion evaluation technique for reinforced concrete structures using magnetic sensors. *Constr. Build. Mater.* **2017**, *135*, 68–75. [CrossRef]
11. Jung, J.-S.; Lee, B.Y.; Lee, K.-S. Experimental study on the structural performance degradation of corrosion-damaged reinforced concrete beams. *Adv. Civ. Eng.* **2019**, *2019*, 1–14. [CrossRef]
12. Fonna, S.; Huzni, S.; Ridha, M.; Ariffin, A.K. Inverse analysis using particle swarm optimization for detecting corrosion profile of rebar in concrete structure. *Eng. Anal. Bound. Elem.* **2013**, *37*, 585–593. [CrossRef]
13. Fonna, S.; Huzni, S.; Ariffin, A.K. Boundary element inverse analysis for rebar corrosion detection: Study on the 2004 tsunami-affected structure in Aceh. *Case Stud. Constr. Mater.* **2018**, *8*, 292–298. [CrossRef]
14. Saleh, E.; Tarawneh, A.; Dwairi, H.; AlHamaydeh, M. Guide to non-destructive concrete strength assessment: Homogeneity tests and sampling plans. *J. Build. Eng.* **2022**, *49*, 104047. [CrossRef]
15. Saleh, E.F.; Tarawneh, A.N.; Katkhuda, H.N. A comprehensive evaluation of existing and new model-identification approaches for non-destructive concrete strength assessment. *Constr. Build. Mater.* **2022**, *334*, 127447. [CrossRef]
16. Amiri, A.S.; Erdogmus, E.; Richter-Egger, D. A Comparison between Ultrasonic Guided Wave Leakage and Half-Cell Potential Methods in Detection of Corrosion in Reinforced Concrete Decks. *Signals* **2021**, *2*, 413–433. [CrossRef]
17. Reddy, V.S.; Naidu, K.S.S.T.; Rao, M.V.S.; Shrihari, S. Electrical Resistivity and Half-Cell Potential Studies to assess organic and inorganic corrosion inhibitors' effectiveness in concrete. *E3S Web Conf.* **2020**, *184*, 1082. [CrossRef]
18. *ASTM C876-15*; Standard Test Method for Corrosion Potentials of Uncoated Reinforcing Steel in Concrete. ASTM International: West Conshohocken, PA, USA, 2016.
19. Adriman, R.; Ibrahim, I.B.M.; Huzni, S.; Fonna, S.; Ariffin, A.K. Improving half-cell potential survey through computational inverse analysis for quantitative corrosion profiling. *Case Stud. Constr. Mater.* **2022**, *16*, E00854. [CrossRef]
20. Gonzalez, J.A.; Miranda, J.M.; Feliu, S. Considerations on reproducibility of potential and corrosion rate measurements in reinforced concrete. *Corros. Sci.* **2004**, *46*, 2467–2485. [CrossRef]
21. Naish, C.C.; Harker, A.; Carney, R.F.A. Concrete inspection: Interpretation of potential and resistivity measurements. Corrosion of reinforcement in concrete. In Proceedings of the Third International Symposium on "Corrosion of Reinforcement in Concrete Construction", Wishaw, UK, 21–24 May 1990; CICC Publisher: UK, 1990.
22. Pradhan, B.; Bhattacharjee, B. Half-cell potential as an indicator of chloride-induced rebar corrosion initiation in RC. *J. Mater. Civ. Eng.* **2009**, *21*, 543–552. [CrossRef]
23. ACI Committee. ACI Committee 211. In *Standard Practice for Selecting Proportions for Normal, Heavyweight and Mass Concrete (Reapproved 2009)*; ACI Committee: Farmington Hills, MI, USA, 2009.
24. ACI Committee. *Building Code Requirements for Structural Concrete (ACI 318-19) and Commentary*; ACI Committee: Farmington Hills, MI, USA, 2019.

coatings

MDPI

Article

Effect of Slurry Coating Modified Methods on Water Absorption of Recycled Coarse Aggregate

Lei Yu [1,*], Sizhong Lv [2], Zhijie Zhao [3] and Zhaolei Liu [3]

[1] Research Institute of Highway Ministry of Transport, Beijing 100088, China
[2] Shandong Highway Incorporated Company, Beijing 250014, China; sdgslsz@126.com
[3] Beijing Xinqiao Technology Development Co., Ltd., Beijing 100088, China; zj.zhao@rioh.cn (Z.Z.); 13910147210@139.com (Z.L.)
* Correspondence: yuleimabel@163.com; Tel.: +86-134-3694-8400

Abstract: Water absorption rate of the recycled coarse aggregate is higher than the natural ones. Until now, there is no test method to record the instantaneous value of the water absorption although this could help us to understand the recycled coarse aggregate better. This paper developed a new device that can record the water absorption continuously and calculate the water absorption rate automatically. The water absorption curve from 0 min to 60 min can be plotted smoothly. The performances of modified recycled coarse aggregate and concretes with recycled coarse aggregate have also been studied. The result shows that the water absorption rate increases fast and about 90% water has been absorbed during the first 10 min. The slurry of silicon nitride with 500 nm particle size can reduce the 10 min water absorption rate of the recycled coarse aggregate from above 4.5% to below 2.5%. A recycled coarse aggregate wrapped with wet slurry is better than the ones wrapped with dry slurry shell for the slump of concrete. Compared to the concrete without any recycled coarse aggregate, the compressive strength and the splitting tensile strength of the concrete with recycled coarse aggregate modified by fresh cement slurry have been increased by more than 20%.

Keywords: recycled coarse aggregate; water absorption; silicon nitride; modified method

1. Introduction

In the reconstruction and extension project of highway, there are concrete structures, such as bridges, the concrete pavement, retaining walls and curbs that need to be torn down. So, a lot of old concrete materials need to be dealt with. If thrown away, they will occupy a quite large field area. Not only will it lead to the waste of resources, but also cause environmental problems. Nowadays, the best way to deal with them is to take advantage of them. Crushing them into coarse aggregate and recycling them into the project would be a good way. Some countries, such as Germany, Japan, Denmark and Korea, have drafted technical regulations to normalize the recycling of concrete [1].

However, the recycled aggregate still has some problems that need to be modified. One is that quite a large amount of mortar sticks to the aggregate. It will cause high water absorption during the mixing process and then lead to the slump of the concrete, decreasing swiftly and obviously [2–5]. Generally, the strength of the recycled coarse aggregates is not as strong as the normal aggregate. It may cause the concrete to have low compressive strength [6–11].

Nowadays, there are two kinds of methods to modify the recycled coarse aggregate. One is breaking the mortar and normal aggregate into two parts. The other one is blocking the pores of the mortar on the aggregate. For the former method, there are three means to achieve this result. One is baking them with a heating device [12,13]. The second is to grind them [14–16]. The third is to steep them with an acid, such as hydrochloric acid [17].

For the method of pore blocking, nanometer materials filling, silicone agent, carbon dioxide curing, mineral admixture slurry wrapped are all efficient methods to block the pores of the mortar on the recycled aggregate [18–22]. Jianhe Xie shows that the pozzolan slurry soaking methods can generally help to strengthen the weak adhered mortar. Forty percent fly ash and silica fume and 3% nano silica fume slurries with 4 h soaking time exhibited a comparable improving performance [23]. Some chemical materials, such as sodium bicarbonate and sodium carbonate, can react with the aqueous calcium hydroxide in the pore of the mortar. The carbonate products often are solid so that they can fill the pore of the mortar [24,25]. The effects of recycled aggregate on the performance of the concrete are also studied. Hailong Wang shows that the permeability of recycled aggregate concrete is much lower than that of the concrete without recycled aggregate due to the on-going hydration and the pozzolanic reaction provided by the phosphorous slag and ground granulated blast-furnace slag additives in the recycled aggregate concrete mixture [26]. Jianhe Xie studied the compressive and flexural behaviours of recycled aggregate concrete modified with silica fume and fibers. The result shows that the coupling effect of silica fume and steel fiber was better than that of the silica fume and polypropylene fibres [27]. Most of the treatments are currently being examined on a small scale under laboratory conditions. How these techniques can be applied to large scale commercial production requires further research and investigation [28].

Considering carbon emission, baking and heating methods will cost a lot of energy. For the acid steeping method, not only will it dissolve the mortar on the aggregate, but also it will cause erosion for the normal aggregate. For the highway construction, the crushed concrete is a kind of waste, and the quantity is very large. Considering the construction cost and convenience, wrapping the admixture slurry outside the aggregate and immersing the aggregate into the silicone agent are two different considerable methods.

The purpose of this paper is to investigate the timely water absorption of the recycled aggregate during 0–60 min and find a properly modified method for decreasing the water absorption of the recycled coarse aggregate. Although the reference shows a newer method to test the water absorption of the recycled coarse aggregate, the data collected are still not continuous [29,30]. Since there is no method to test the timely water absorption of the aggregate before, a new method has to be invented in this paper. The size of the particle can affect the amount of its slurry absorbed by the recycled coarse aggregate [31]. The wave method and soaking the aggregate in acid are not all convenient methods for field construction [32]. Considering the particle size in the slurry and the construction convenience, the slurry made by cement, silica fume, silicon nitride is used for wrapping the recycled aggregate to decrease the 0–60 min water absorption of the recycled coarse aggregate. To know the effect of the modified recycled aggregate on the concrete, the concretes with recycled coarse aggregate wrapped by wet slurry or by dry slurry shell are made to distinguish the slump and mechanical performance. By testing the slump and mechanical performance of the concrete with modified recycled coarse aggregate, the effects of the modified method to the concrete can be clear. To save cost for construction, the mixing way of the aggregate modified by cement slurry is selected by contrast experiment. Finally, this paper tries to find a mixing method with a high performance to cost ratio for the mixing progress during construction.

2. Materials and Methods

2.1. Materials

2.1.1. Recycled Coarse Aggregate

In order to obtain the common rules of the water absorption of recycled coarse aggregate, two different origin recycled coarse aggregates are prepared. One is from the slabs of the bridges of a highway which is located in Jinan City, Shandong province, China. The compressive strength range of the concrete from these bridges is from 5800 psi to 7250 psi. The other is from the cement concrete pavements located in the same area. The compressive strength range of the concrete from the cement concrete pavements is from 4350 psi to 5800 psi. The aggregate from the bridges is numbered 1# and the one from the cement concrete pavement is numbered 2#. Both 1# and 2# include two particle sizes. They are

5–10 mm and 10–20 mm. The properties of these two kinds of aggregates are listed in Table 1. The apparent density and the 24 h water absorption are tested based on the ASTM C127-2015 [33]. The crushing value of the aggregate is tested based on the Chinese national standard, for which the number is GB/T 14685-2011 [34]. The chlorine ion content is tested based on the Chinese national standard, for which the number is JGJ/T 322-2013 [35].

Table 1. Properties of the aggregates.

ID	Range of Particle Size (mm)	Apparent Density (kg/m^3)	Crushing Value (%)	24 h Water Absorption (%)	Chlorine Ion Content (%)
1#	5–10	2671	21.5	6.11	0.0475
	10–20	2676		5.34	
2#	5–10	2689	29.5	6.83	0.0037
	10–20	2687		5.17	

2.1.2. Materials for Mixing Slurry to Modify the Recycled Aggregate

Four kinds of materials are used for modifying the water absorption of the aggregate by coating them. They are cement (C), silica fume (Si), nanometer silicon nitride (SiN) and silicone hydrophobic agent (SiH-agent). About the silica fume, there are two different diameter particles of 150 nm and 300 nm. The material ID can be defined as Si-150 nm, Si-300 nm. For the nanometer silicon nitride, there are also two kinds of particles such as 20 nm and 500 nm. The material IDs can be defined as SiN-20 nm, SiN-500 nm. The properties of the cement, silica fume and silicon nitride are listed in Table 2. The density and the specific surface area are tested separately based on the standard of ASTM C188-2009 and ASTM C204-2016 [36,37]. The properties of the silicone hydrophobic agent are listed in Table 3. The specification of the silicone hydrophobic agent is shown in the link [38]. Hydroxy Propyl Methyl Cellulose is used to increase the consistency of the slurry. The properties of the cellulose are listed in Table 4. The specification of the cellulose is shown in the link [39].

Table 2. Properties of the cement, silica fume and silicon nitride.

ID	Particle Size	Density (kg/m^3)	Specific Surface Area (m^2/kg)
C	/	3100	370
Si-150 nm	150 nm	300	28,000
Si-300 nm	300 nm	700	20,000
SiN-20 nm	20 nm	90	90,000
SiN-500 nm	500 nm	230	35,000

Table 3. Properties of the silicone hydrophobic agent.

Style	PH Value	Solid Content (%)	Density (kg/m^3)	Recommended Mixing Amount (%) of Cement Mass
FS-170	7.0	20	1015	1

Table 4. Properties of the cellulose.

Style	Fineness (μm)	Density (kg/m^3)	Color
GLHPMC01	150	500	white

2.1.3. Common Materials for Concrete

Commercial cement (grade 42.5) was used in the experiments. The properties of the cement are listed in Table 5. A polycarboxylate-based superplasticizer was used to guarantee that fresh concrete has a considerable slump. The mixing content is 1% of the cement amount. Natural sand was used as fine aggregate, and the measured value of the fineness modulus is 2.78. The cumulative screening percentage of the natural sand is listed in Table 6. Natural gravel with a particle size of (5–10) mm and (10–20) mm are used as normal coarse aggregate. The origin of the natural gravel is Shandong province in China. The 24 h water absorption, the crushing value and the apparent density of the normal coarse aggregate is 0.45%, 10.5% and 2730 kg/m^3, respectively. The properties of the cement are tested based on the Chinese standard GB 175-2020 [40]. The cumulative screening percentage of the natural sand is tested based on the Chinese standard GB/T 14684-2011 [41].

Table 5. The properties of cement.

Fineness (80 μm Square Hole Sieve) (%)	Water Requirement for Normal Consistency (%)	Setting Time (h: min)		Compressive Strength (MPa)		Splitting Tensile Strength (MPa)		Stability (Boil)
		Initial	Final	3 d	28 d	3 d	28 d	
3.1	27.6	2:59	3:59	29.8	49.9	6.7	9.3	qualified

Table 6. The cumulative screening percentage of the natural sand.

Sieve Size (mm)	4.75	2.36	1.18	0.6	0.3	0.15	0.075	<0.075
Cumulative screening percentage	11.62	32.2	43.58	55.6	79	93.3	99.9	100

2.2. Test Measurements

2.2.1. Water Absorption Measurements

A common method to detect the water absorption of the aggregate is soaking them in water and then dry them with a moist towel until the saturated surface reaches dry status according to ASTM C127-2015 [33]. The progress is obviously discontinuous. In order to obtain continuous data of the water absorption and plot the whole curve of the aggregate smoothly, I introduced a new method. The details of the measurement are shown as follows.

Take 5000 g coarse recycled aggregate and sieve them by the 4.75 mm sieve with square pores. Wash the rest of the coarse recycled aggregate and dry them by drying oven for 48 h. In order to avoid thermal decomposition in the mortar, set the drying oven temperature as 78 ± 3 °C instead of 105 ± 5 °C. Take the dry aggregate out of oven and static them cool down in the dryer with temperature 20 ± 2 °C. Quarter them and take out double sections. Weigh 1000 g aggregate from each section preparing for the next test. The water absorption testing apparatus is displayed in Figure 1.

First of all, put the net basket on the hook. Adjust the temperature of the water between 15 and 25 °C. Then, pour water into the overflow tank until the water begins flowing from the overflow port. Shut off the overflow port. Adjust the electronic hydrostatical balance to the number zero. Weigh m gram (m = 1000 ± 5 g) prepared dry coarse aggregate and put them into the net basket as soon as possible and not exceeding 30 s. Beat the net basket so that the air can expel from the net basket. Read the number of the 30th second on the balance as the initial quality of the coarse aggregate in water. Read the number of the 24th hour on the balance as the final quality of the coarse aggregate in water and denote it as m′. The principle of reading the number between the initial number and the final one is shown in Table 7. Considering the quick rate at which water absorption happens during the early stage, to draw a smooth curve, the frequency of reading the number in the early stage must be much higher than the late stage. The quality of the aggregate in water between the 30 s and 24 h is recorded as m″. After 24 h, take the coarse aggregate out of the water and dry them with a wet cotton cloth until the surface of the aggregate has no obvious water. Weigh them and record as m‴. The water absorption of the coarse aggregate can be calculated as Equation (1).

$$\omega = \frac{m'' + m''' - m' - m}{m} \times 100 \quad (1)$$

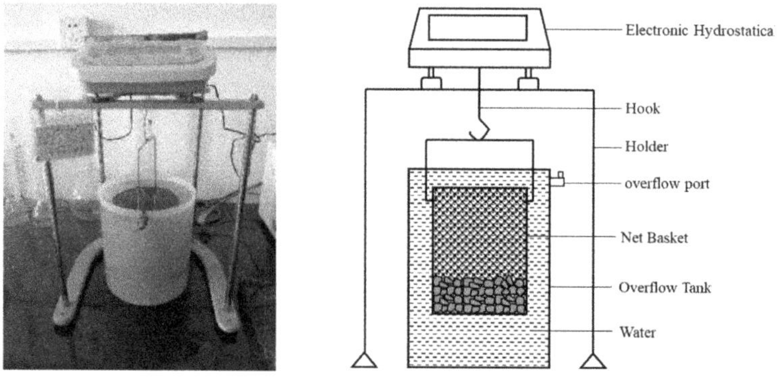

(a) (b)

Figure 1. Water absorption testing apparatus: (**a**) Picture; (**b**) Schema.

Table 7. The principle for recording the number.

Test Time	30 s–10 min	10 min–20 min	20 min–60 min	30 min–1 h	1 h–6 h
Frequency of record number	30 s	1 min	5 min	10 min	1 h

2.2.2. Coarse Aggregate Modified

In order to wrap the coarse aggregate with the cement, silica fume and nanometer silicon nitride, the slurries of these powder have been made.

Reduce the aggregate by quartering the method to about 500 g and set aside into two parts. First of all, prepare the aggregate just like in Section 2.2.1. Then, put the cool, dry, prepared aggregate into the slurry, mix them until the particles are covered with slurry. Static the mixture for 2 h. For cement slurry, it should be stirred every 10 min to avoid congealing during the static time. After 2 h, sieve the mixture with the 2.36 mm sieve until that there is no obvious slurry dropping. Spread the aggregate on a pallet and carry them into the dry oven. The steps to dry the aggregate and the measurement of water absorption are the same as Section 2.2.1. The appearance of the aggregate covered with different materials is shown as Figure 2.

2.2.3. Performance Test of Concretes Made by the Recycled Coarse Aggregate

In order to distinguish the difference of aggregates modified by above methods, the concretes are prepared respectively by the modified aggregates above. The slumps of initial, 5 min, 10 min, 15 min have been tested according to ASTM C143/C143M-12 to observe the effects that the different aggregate modified methods have compacts on the concrete slumps [42]. The 28-d compress strength have also been tested according to ISO 4012-1978 [43] to verify that the modified methods cannot reduce the strength of the concrete.

Figure 2. The appearance of the aggregate wrapped with different material slurry: (**a**) Wrapped with cement slurry; (**b**) Wrapped with silica fume slurry (left: 150 nm, right: 300 nm); (**c**) Wrapped with silicon nitride slurry (left: 20 nm, right: 500 nm); (**d**) Wrapped with silicone hydrophobic agent.

2.2.4. The Mixing Procedure of the Concrete with the Dry Modified Aggregate and the Wet One

Considering concrete production in practical engineering, the mixing technology will be different between the dry modified aggregate and the wet one. If using dry modified aggregate, there may need to be a slurry pool at the construction site. First of all, prepare the slurry mixed by the modified material in the slurry pool, and then put the recycled coarse aggregate in the pool, wait for a certain time more than 1 h, fish out the aggregate, dry them under the natural sun or with other methods. Then, the dry modified recycled coarse aggregate can be used to make concrete.

If using the wet modified aggregate, the steps of treating the recycled coarse aggregate above can be simplified due to the dry step can be cut off.

2.2.5. Mixing Procedure of the Concrete with the Cement Slurry Wrapped Aggregate

There are two different mixing styles. One is preparing the wet modified aggregate with cement slurry as in Section 2.2.4. Then mix the prepared wet modified aggregate into the blender. The other method is mixing additional cement and water that the coarse aggregate wrapping needed firstly in the blender and then mixing all of the concrete raw materials including the recycled coarse aggregate and the normal coarse aggregate together at the same time. The contrast experiment of these two methods has been completed. In the contrast experiment, the 1# recycled coarse aggregates with (5–10) mm and (10–15) mm were used. The C-W group was made by the former method, but the C-T group was made by the latter method.

The quantity of the additional cement slurry which includes cement and water should be calculated with Equations (2) and (3), respectively.

$$m_c^{C-T} = (m_1\omega_1 + m_2\omega_2) \times \frac{m_c}{m_c + m_w} \tag{2}$$

$$m_w^{C-T} = (m_1\omega_1 + m_2\omega_2) \times \frac{m_w}{m_c + m_w} \tag{3}$$

In these two formulas, m_c^{C-T}—the extra cement quantity; m_w^{C-T}—the extra water quantity; m_1—the quantity of the unit recycled coarse aggregate with (5–10) mm; m_2—the quantity of the unit recycled coarse aggregate with (10–20) mm particle size; m_c—the quantity of cement in the original mix proportion; m_w—the quantity of water in the original mix proportion; the unit of all weight parameters is kilogram; ω_1—the water absorption of the recycled coarse aggregate with (5–10) mm particle size; ω_2—the water absorption of the recycled coarse aggregate with (10–20) mm particle size; the unit of the water absorption is %.

2.3. *The Mix Proportion of Slurry and Concrete*

2.3.1. The Mix Proportion of Slurry

The mix proportions of the slurry used for wrapping the recycled aggregate are listed in Table 8.

Table 8. The mix proportion of the different slurry (kg/m^3).

Cement	Silica Fume		Nanometer Silicon Nitride		Water	Cellulose
	150 nm	300 nm	20 nm	500 nm		
500	/	/	/	/	264	5
/	150	/	/	/	300	2
/	/	150	/	/	300	2
/	/	/	30	/	600	/
/	/	/	/	100	130	/

2.3.2. The Mix Proportion of Cement

According to the experiment design in Section 2.2.3 and calculation method in Section 2.2.5, a certain amount of normal coarse aggregate will be replaced by the 1# recycled coarse aggregate. The C30 concrete was made to test the slump and mechanical performance. The details of the mix proportions of concrete with different aggregates are listed in Table 9. Group None is the treatment group without any recycled coarse aggregate. Groups with D letter concretes include a half of recycled coarse aggregate covered with dry slurry shell. The serial number of groups with D letter will vary along with the different slurry shell, as shown in Table 9. For example, if the recycled coarse aggregate is covered with cement slurry shell, then it is numbered C-D (cement-dry). Relatively, Group W is on behalf of the concretes with the wet slurry wrapping on the recycled coarse aggregate.

The numbers will also vary along with the different wet slurry materials. For example, if the recycled coarse aggregate is covered with fresh cement slurry, then it can be numbered C-W (cement-wet). Group C-T has been introduced in Section 2.2.5.

Table 9. Mix proportions of concrete with different aggregates (kg/m^3).

No.	Cement	Water	Sand	Coarse Aggregate			
				(5–10) mm		(10–20) mm	
				Normal	Recycled	Normal	Recycled
None	332	175	860	420	0	632	0
C-D	332	175	860	210	210	316	316
C-W	313	165	860	210	222	316	333
C-T	351	185	860	210	210	316	316
Si-150 nm-D	332	175	860	210	210	316	316
Si-150 nm-W	332	175	860	210	210	316	316
SiN-20 nm-D	332	175	860	210	210	316	316
SiN-20 nm-W	332	175	860	210	210	316	316

From Table 9, C-D, C-W and C-T groups have different unit weights of materials; this is because the recycled coarse aggregate adsorbs fresh cement slurry. The fresh cement slurry will enhance the compressive strength of the concrete. It will lead to fuzziness of whether the compressive strength is due to the modified aggregate or the spare fresh cement slurry. So, the cement and the water from the fresh cement slurry which is absorbed by the recycled coarse aggregate must be deducted. The amount of the deducting weight of the cement and water are calculated by the formula 2 and 3 in Section 2.2.5. In a similar way, considering the comparability between group C-W and C-T, the group C-T has to add equal cement and water so that these two groups have the same water and cement ratio.

3. Results

3.1. Water Absorption of Recycled Coarse Aggregate

In order to obtain common rules for water absorption of recycled coarse aggregate, two different origin recycled coarse aggregates are prepared. The water absorption of these two kinds of recycled coarse aggregate from 0 min to 60 min are displayed in Figure 3.

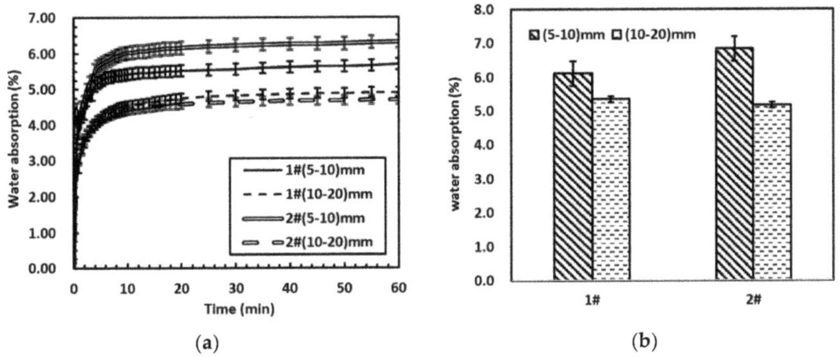

Figure 3. Water absorption of the different origin recycled coarse aggregates, (**a**) 0 min–60 min, (**b**) 24 h.

From Figure 3a, it can be seen that the curve tend of the two kinds of recycled coarse aggregates are almost the same as each other. During the first 10 min, water absorption is quick. After 10 min, the rate of the water absorption becomes smooth. For different recycled coarse aggregate, about 90% of water has been absorbed in the first 10 min. From Figure 3, it can also be seen that different origin recycled coarse aggregate may have different water absorption characteristics. The water absorption of the aggregate with (5–10) mm particle size is absolutely more than the one with (10–20) mm particle size. This may be because the content of mortar sticking on or mixing in the (5–10) mm particles is more than that sticking on the (10–20) mm aggregate. The mortar has stronger water absorption ability than the natural aggregate. The more content of mortar in the aggregate, the stronger water absorption ability it will have. From Figure 3b, it can be seen that the rule is the same as the 60th minute. No matter 1# or 2#, the water absorption of the (5–10) mm aggregate is larger than that of the (10–20) mm ones. The water absorption of 2# aggregate with (5–10) mm particle size is 6.83% which is a little bigger than the one of 1# aggregate which is 6.11. The opposite phenomenon, that 1# water absorption is a little bigger than 2# water absorption, appears with the (10–20) mm particles. Exploring the reason, it may be due to the different origin of the aggregates. The 1# comes from the slab of the bridge. The 2# comes from the pavement. In China, there is a common rule that the concrete grade of bridge slab is larger than that of the pavement. In general, the concrete grade range of the bridge slab can be C45-C50 but the one of pavement can be C25-C30. The number of pores is the main reason for the water absorption in the concrete. So, the water absorption of 2# with (10–20) mm particle size is larger than the one of 1#. Why does the water absorption in (5–10) mm have the opposite phenomenon? This may be because the content of mortar in 1# is more than that in 2#. Due to the mechanical disruption during production process, the mortar in the concrete is broken into small particles with (5–10) mm but not (10–20) mm. The bridge slab with higher strength, smaller water cement ratio and more cement content has more mortar than the concrete pavement with lower strength, higher water cement ratio and low cement content.

3.2. Water Absorption of Modified Recycled Coarse Aggregate

To find effective methods of modified recycled coarse aggregate so that reduce the water absorption, four kinds of materials are selected to mix them with water into slurry for wrapping the recycled coarse aggregate. Make the 1# aggregate as a representative. The water absorption of modified recycled coarse aggregates is displayed in Figure 4.

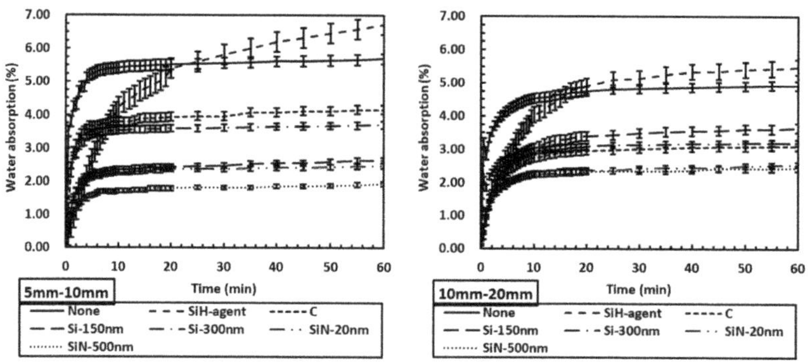

(a) (b)

Figure 4. The water absorption of modified recycled coarse aggregate, (**a**) (5–10) mm, (**b**) (10–20) mm.

First, from Figure 4, after wrapping the slurry shell, no matter the (5–10) mm recycled coarse aggregate or the (10–20) mm ones, the (0–10) min water absorption of the recycled coarse aggregate is significantly reduced. Noteworthy, the SiH-agent can slow down the first 10 min water absorption rate of the recycled coarse aggregate and can prolong the sum of the first 10 min water absorption to 20 min. However, it can increase the sum water absorption at last. So, it may not be a good modified material. C, Si, SiN can change the water absorption curve and reduce the sum water absorption absolutely. However, they also cannot decrease the first 10 min water absorption rate. SiN material is the best one for reducing the water absorption compared to C and Si.

From Figure 4a, for the modified recycled coarse aggregate with (5–10) mm particle size, the water absorption is ordered as SiN-500 nm > SiN-20 nm > Si-150 nm > Si-300 nm > C. Comparing to the unmodified recycled coarse aggregate, by the end of 60 min, the sum water absorption can be decreased to 1.93%, 2.47%, 2.64%, 3.70% and 4.16%. The decreasing amplitude reached 66.1%, 56.6%, 53.7%, 35.0% and 27.0%, respectively. From Figure 4b, for the modified recycled coarse aggregate with (10–20) mm particle size, the water absorption is ordered as SiN-500 nm > SiN-20 nm > C > Si-300 nm > Si-150 nm. The 60 min sum water absorption can be decreased to 2.43%, 2.54%, 3.08%, 3.19% and 3.65%. The decreasing amplitude reached 50.5%, 48.3%, 37.3%, 35.0% and 25.8%, respectively. From the date above, it can be seen that the same modified material has different effects on the different recycled coarse aggregate. Using the same modified material, the water absorption reducing amplitude of the (5–10) mm particle size recycled coarse aggregate is more than the one with (10–20) mm particle size.

To analyze the reason for the different modified effects from C, Si, SiN materials, the particle size and the chemical properties may be the main two reasons. As we all know, there are a lot of pores in recycled coarse aggregate due to the existence of the sticking mortar. According to the previous study, the pore diameter of the mortar ranges from nanometer size to millimeter size [44]. However, the pore which can determine the water absorbing ability of the mortar is the capillary pore. The proportion of the capillary pore is about 0–40% in cement-based mortars, and its size is among (100–1000) nm [45]. In order to decrease the water absorption of the recycled coarse aggregate, the particle size which is used for blocking the capillary pore has to be smaller than 50 μm. So, the materials selected in the experiment are all suitable. Ordinarily, the smaller the particle is, the easier it can permeate into and block the pores in mortar. From Figure 4, this regularity is not remarkable. The water absorption of the recycled coarse aggregate modified with SiN-500 nm is less than the one modified with SiN-20 nm. Based on Figure 4b, this regularity also happens. The water absorption of the recycled coarse aggregate modified with Si-300 nm is less than the one modified with Si-150 nm. So, there may be other reasons leading to this result that need to be studied, such as detecting the modified aggregate pore distribution to see the amount of the rest capillary or observing the interspace of the capillary pore to see how much the interspace of the pore is occupied by the modified particle. Maybe these methods can explain the phenomenon that the smaller particle has weaker blocking ability.

The chemical properties of these particles can also have an effect on the water absorption of the recycled coarse aggregate. The wrapping materials can be separated into three kinds. One is pure sluggish material such as the silicon nitride. The second is half activity material such as the silica fume. The third one is activity material such as cement. The sluggishness material can not react with the water. It is only used as a filling material to fill the capillary pore of the mortar sticking on the recycled coarse aggregate. The half activity material can react with the water, not immediately, but sometime later. For example, the silica fume, if you add it to the cement base slurry, the silica fume can react with water slowly after the 28-d curing period. So, during the steeping process of recycled coarse aggregate in the slurry, the silica fume can absorb more water than the sluggish material. The water of the recycled coarse aggregate modified with silica fume slurry is composed of two parts, one part is from the pores which are not blocked by the silica fume slurry and the other part is from the reacting between the silica fume and the water. The active material can react with water almost immediately as long as it comes across the water. For example, the cement, the cement-based slurry will be wrapped on the recycled coarse aggregate and then dried. When testing the water absorption of the modified recycled coarse aggregate, the water that the aggregate absorbed can also be separated into two parts similar to the silica fume slurry. Because the water that the cement hydrate needs is much more than the silica fume hydrate, the water absorption of the recycled coarse aggregate modified by the cement slurry will be more than the ones modified by the silica fume slurry.

What is more, the dry cement slurry shell on the recycled coarse aggregate will continue to hydrate when it comes across the water. So, after 60 min, the cement will continue absorbing water to support the cement hydrate. The 24 h water absorption has been tested and displayed in Figure 5 to verify the final water absorption situation.

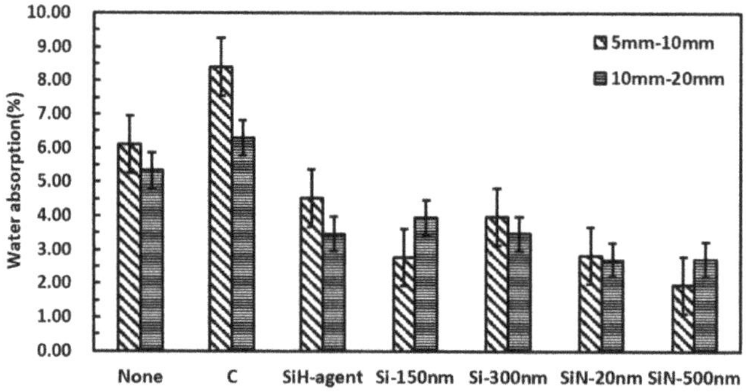

Figure 5. The 24 h water absorption of recycled coarse aggregate modified by different materials.

From Figure 5, it can be seen that almost every material can reduce the 24 h water absorption of the recycled coarse aggregate excluding the cement. Obviously, the water absorption of the aggregate modified by cement slurry is more than other materials and even more than the original recycled aggregate. This special phenomenon is due to the continuous cement hydrate reaction. For the (5–10) mm recycled coarse aggregate, the 24 h water absorption is ordered as C > None > SiH-agent > Si-300 nm > SiN-20 nm > Si-150 nm > SiN-500 nm. The water absorption is separately listed as 8.40%, 6.11%, 4.52%, 3.97%, 2.84%, 2.78%, 1.95%. The best modified material is SiN-500 nm, and it can reduce the water absorption of the recycled coarse aggregate by more than 60%. However, the price of the silicon nitride is much higher than the silica fume. The Si-150 nm is the second most excellent material. It can reduce the water absorption of the recycled coarse aggregate from 6.11% to 2.78%. The reduction in amplitude is about 54.5%. A little less than the SiN-500 nm. However, the price of the Si-150 nm is much less than the SiN-500 nm. If there is a large amount of the recycled coarse aggregate with (5–10) mm particle size, the Si-150 nm may be the best cost performance ratio material used for modifying these recycled coarse aggregate. For the (10–20) mm recycled coarse aggregate, the 24 h water absorption is ordered as C > None > Si-150 nm > Si-300 nm > SiH-agent > SiN-500 nm > SiN-20 nm. The water absorption is listed separately as 6.29%, 5.34%, 3.95%, 3.47%, 3.46%, 2.72%, 2.70%. For the (10–20) mm recycled coarse aggregate, the best modified material is SiN-20 nm and it can reduce the water absorption of the recycled coarse aggregate about 50%. Although the SiN material is better than other materials, the high price may limit its use for engineering.

From Sections 3.1 and 3.2, no matter the original recycled coarse aggregate or the modified ones, the water absorption of the recycled coarse aggregate is still higher than the natural coarse aggregate with water absorption under 2% according to the Chinese standard GB/T 14685-2011.

3.3. *Concrete Performance Made by Modified Recycled Coarse Aggregate*

Comparing the different effects on the concrete coming from the aggregates with different modified methods and different dry and wet conditions, the cement, silica fume with particle size 150 nm and the silicon nitride with 20 nm are selected. The dry modified recycled coarse aggregates are prepared by the method of Section 2.2.2. Number them as C-D, Si-150 nm-D and SiN-20 nm-D. The wet ones are also prepared following the Section 2.2.2 but without dry steps. Number them as C-W, Si-150 nm-W and SiN-20 nm-W.

Number the unmodified recycled coarse aggregate with None. The slump, compressive strength, splitting strength have been tested. The results are displayed in Figures 6–8. The mix proportion and the percentage of the recycled coarse aggregate are the same among these concretes. The details are shown in Table 9. The raw materials are all the same. The only difference is the kind of modified materials.

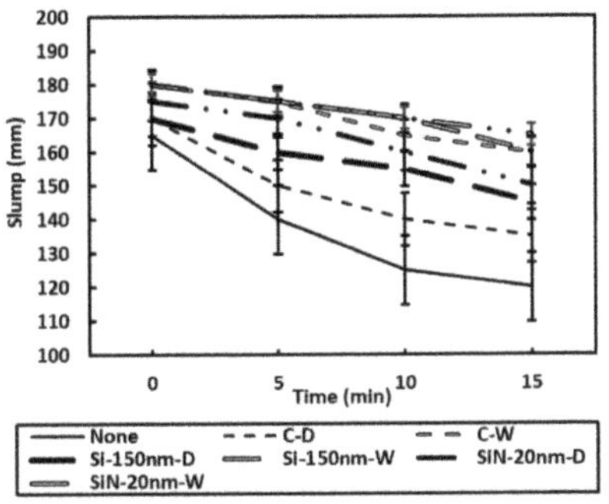

Figure 6. The slumps of concrete with different modified materials.

Figure 6 shows the slump change of the concretes with different modified materials and different dry or wet conditions among 0–15 min. First, from Figure 6, it can be seen that the slump will decrease with the time flow. However, the decreasing amplitude is different. The most important thing is that the wet modified recycled coarse aggregate is better than the dry ones. Not only is the initial slump of the concrete with the wet modified recycled coarse aggregate more than that of the concrete with the dry ones, but also the whole slump curve of the concrete with the wet modified recycled coarse aggregate is all on the top of the dry ones. From Figure 6, the initial slump can be ordered as SiN-20 nm-W (180 mm) = Si-150 nm-W = C-W > SiN-20 nm-D (175 mm) > Si-150 nm-D (170 mm) = C-D > None (165 mm). The 15th min slump can be ordered as SiN-20 nm-W (165 mm) > Si-150 nm-W (160 mm) = C-W > SiN-20 nm-D (150 mm) > Si-150 nm-D (145 mm) > C-D (135 mm) > None (120 mm). Using the initial slump minus the 15th min slump, the slump decreasing amplitude of the concrete with different modified materials between 0 min to 15 min is plot in Figure 7.

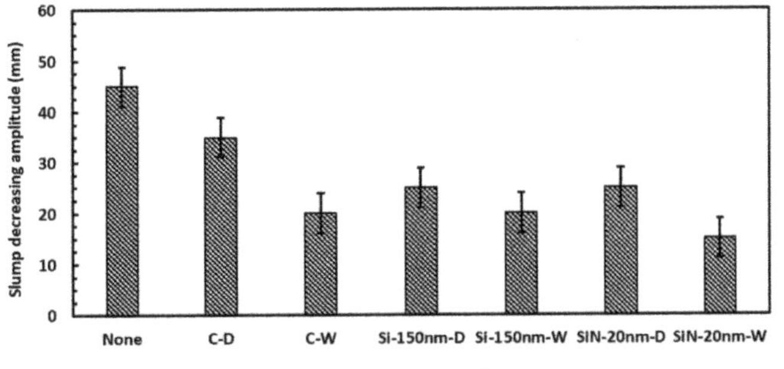

Figure 7. The slump decreasing amplitude of the concretes.

From Figure 7, the slump decreasing amplitude is ordered as SiN-20 nm-W (15 mm) > Si-150 nm-W (20 mm) = C-W > SiN-20 nm-D (25 mm) = Si-150 nm-D > C-D (35 mm) > None (45 mm). Actually, the less of the decreasing amplitude of the concrete is, the better of the concrete construction performance is. So, for keeping the slump of the concrete, the wet modified recycled coarse aggregate may be better than the dry ones. Considering the cost performance ratio, the cement is a more proper material than the silica fume and silicon nitride. So, during the real construction, using the cement slurry as the modified materials is a good selection.

The compressive strength and splitting strength of the concrete made by different modified materials are shown in Figure 8.

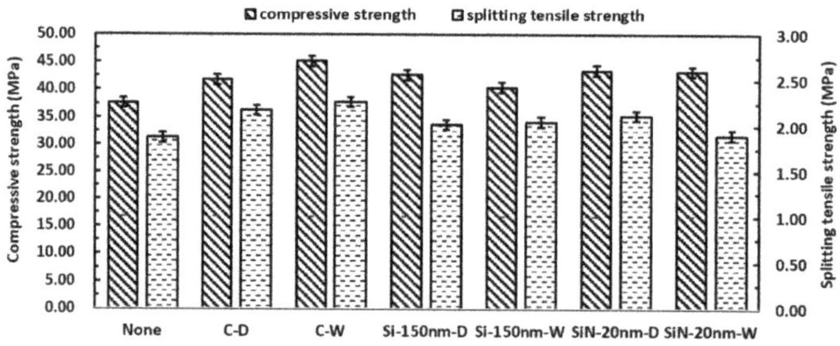

Figure 8. The mechanical property of the concrete made by different modified materials.

From Figure 8, the compressive strength of the concrete with different modified recycled coarse aggregate is plot on left axis and the splitting tensile strength is plot on the right one. Compared to the None group, the compressive strength and splitting tensile strength of all other groups are all increased. This phenomenon proves that the modified recycled coarse aggregate can enhance the mechanical property of the concrete.

From Figure 8, the compressive strength can be ordered as C-W > SiN-20 nm-D > SiN-20 nm-W > Si-150 nm-D > C-D > Si-150 nm-W. The increasing amplitude is listed separately as 20.12%, 15.60%, 15.16%, 13.48%, 11.08%, 7.45%. Unlike the slump rules, the concrete with wet modified recycled coarse aggregate is not better than the dry ones. The compressive strength of the concrete with SiN-20 nm-D modified the aggregate is more than the one with the SiN-20 nm-W modified the aggregate. The same situation also happens on the Silica fume and Cement. This phenomenon may be due to the aggregate with dry slurry shell wrapped can absorb more water than the aggregate with wet slurry on them. So, the water amount may decrease indirectly, and the water cement ratio is decreased at the same time. The smaller water cement ratio can lead to more compressive strength. The most important thing, from Figure 8, it also can be seen that the C-W group is the best one among these groups. From Figure 8, the splitting tensile strength can be ordered as C-W > C-D > SiN-20 nm-D > Si-150 nm-W > Si-150 nm-D > SiN-20 nm-W. The increasing amplitude is listed separately as 20.83%, 16.20%, 12.86%, 9.30%, 7.72%, 1.36%. So, the C-W is also the best group for the tensile strength. From the slump and mechanical performance analysis above, it can be seen that the C-W is definitely a good method for modifying the recycled coarse aggregate. Not only does it have good slump maintain performance, good mechanical performance, but also it is cheaper than the silica fume and silicon nitride.

3.4. The Effect of the Mixing Style of the Aggregate with Cement Slurry Wrapped

From the above study results, the C-W is a good way to modify the recycled coarse aggregate. However, considering the cost and convenience, setting a slurry pool on the construction site for pre-absorb slurry is not a high performance-cost-ratio thing. So, the study tries to find a new convenient way to wrap the recycled coarse aggregate with cement slurry.

Actually, the mechanism of modified recycled coarse aggregate has a lower water absorption rate, mainly due to the cement slurry blocking the pore of the mortar on the recycled aggregate. The slump of the concrete with recycled coarse aggregate decreased mainly due to the pores in the mortar on the aggregate. So, if there is enough cement slurry that the recycled coarse aggregate can absorb sufficiently, the recycled concrete slump decreasing can be resolved. Based on this method, this study decided to mix enough cement slurry into the concrete during the mixing process in order to supply enough cement slurry for the un-wrapped recycled coarse aggregate.

The slump, mechanical performance of the concrete has been tested. The results are displayed in Figures 9 and 10.

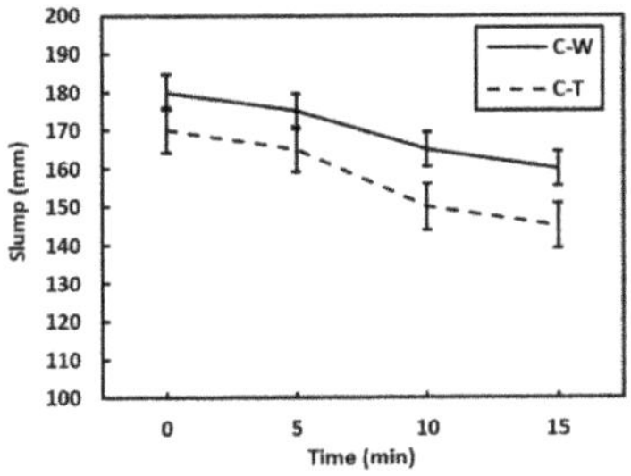

Figure 9. The slump of concrete (C-W group vs. C-T group).

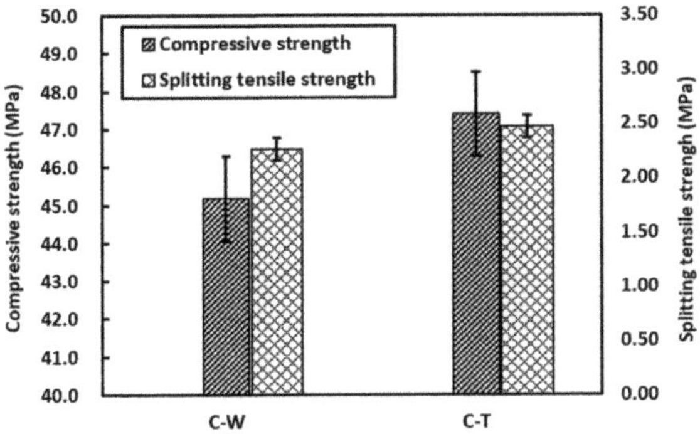

Figure 10. The mechanical performance of concrete (C-W group vs. C-T group).

From Figure 9, the slump of C-W group is more than the C-T group. For the initial slump of concrete, the C-W group which is 180 mm is also more than the C-T group which is 170 mm. For the slump decreasing amplitude, the C-W group which is 20 mm is less than the C-T group which is 25 mm. So, from the aspect of slump, the C-W is better than the C-T. This may be due to the C-W group aggregate being wrapped with cement slurry before being put into the blender. That means the pores of the mortar on the recycled coarse aggregate have been blocked in advance. So, the water absorption of the recycled coarse aggregate has been decreased before being put into the blender. During the mixing process in the blender, the wrapped recycled coarse aggregate has lower water absorption performance than the aggregate of the C-T group which has not been treated in advance. Meanwhile, the good thing is that the slump and the slump decreasing amplitude of concrete with C-T group aggregate are all still satisfied the requirement of the concrete construction.

From Figure 10, it can be seen that the compressive strength of C-W group which is 45.2 MPa is a little less than that of the C-T group which is 47.39 MPa. The splitting tensile strength shows the same rule as the compressive strength. So, the mechanical performance of concrete has different rules relating to the slump of concrete. The reason may be the extra cement slurry which is pre-mixing in the blender during the mixing process. The extra cement slurry takes more cement into the mixture. The more cement content the concrete has, the better compressive strength it will have.

From the slump and mechanical performance of the concrete, it can be seen that the C-T group can also satisfy the construction need as well as the C-W group. This result went quite well. That means there is no need to set up a slurry pool on a construction site. Good performance recycled concrete can be obtained by adding extra cement slurry directly.

4. Discussion

The test method of timely water absorption of the recycled coarse aggregate was invented first so that the whole curving of the water absorption of the recycled coarse aggregate can be plotted. Different recycled coarse aggregates have different water absorption curves. In the future, it may be a good way to distinguish them by the curves in concrete engineering.

Although the water absorption of the recycled coarse aggregate can be decreased by nano material slurry, some mechanisms still need to be studied. For example, how does the slurry flow in the tube of the pores? What is the diameter range of the particle which can block off the capillary pore maximum? Whether the status of the nano particle in the pores is stable enough to so that the cement concrete cannot be affected by the free nano materials.

The micro pore structure should be studied in the future paper. For example, comparing to recycled coarse aggregate without blocking how the distribution of the pore structure of the cement concrete changes. Maybe these studies can explain why the water absorption decreased by the nano material slurry.

Although the water absorption rate of the recycled coarse aggregate has been decreased by the cover shell of the slurry, more than 2% water absorption rate is still larger than natural aggregate with below 1% water absorption rate. So, to produce the cement concrete with recycled coarse aggregate, other methods such as using water reducer agent, optimizing the grading of the aggregate are all needed.

The durability of the cement concrete with recycled coarse aggregate needs to be studied especially the anti-frozen performance. This paper only studied the slump and compressive strength of the cement concrete with recycled coarse aggregate.

5. Conclusions

To obtain the water absorption curve of the recycled coarse aggregate from 0 min to 60 min, a new device has been invented. a new test method has been explored. Water absorption calculated for any time has been given. To reduce the water absorption of the recycled coarse aggregate, four kinds of materials are used such as cement, silica fume, nanometer silicon nitride and silicone hydrophobic agent. To find the effects of the recycled coarse aggregate on the concrete, the slump and mechanical performance of concretes have been tested. To find the mixing order effects, two kinds of mixing order have been studied. The results can be summarized as follows.

(1) A new device has been invented to record the water absorption continuously and calculate the water absorption rate automatically. The device consists of hardware and software. The hardware is shown in Figure 1. The software is designed to record and calculate the test data according to Equation (1). The test procedure has been given in Section 2.2.1. With this device, the water absorption curve can be plotted smoothly owing to the water absorption at any time can be recorded.

(2) During the first 10 min, the water absorption rate of the recycled coarse aggregates increases quickly. After 10 min, it becomes smoother. For different recycled coarse aggregate, about 90% amount of water has been absorbed in the former 10 min. The water absorption of the aggregate with (5–10) mm particle size is absolutely more than the one with (10–20) mm particle size.

(3) The 0 min–10 min water absorption of the modified recycled coarse aggregate is significantly reduced. C, Si, SiN can absolutely reduce the water absorption amount, and SiN is the best one. For the SiN-50 nm, it can reduce the first 10 min and the 60 min water absorption from above 4.5% to below 2.5%. It can also reduce the 24 h water absorption from above 5.3% to below 2.8%. For the amplitude of the 60 min water absorption, it is more obvious in the (5–10) mm particle size aggregate than in the (10–20) mm ones.

(4) The water absorption of the aggregate modified by cement slurry is more than other materials and even more than the original recycled aggregate. This special phenomenon is due to the continuous cement hydrate reaction. After 60 min, the dry cement shell on the aggregate will continue absorbing the water to support the cement hydrate performance.

(5) For the slump decreasing amplitude, the wet modified recycled coarse aggregate is generally better than the dry ones. The C-W group has the best strength among all of the groups. Comparing the None group, both the compressive strength and the splitting tensile strength of the C-W group concrete have been increased more than 20%.

(6) The slump of C-W group is better than the C-T group. The mechanical performance is opposite. The strength of the C-T group is larger than that of the C-W group. The reason may be the extra cement slurry which is pre-mixing in the blender leading to better strength. Good performance recycled concrete can be obtained by adding extra cement slurry to the blender directly.

6. Patents

A kind of test method and facility for testing the water absorption of the recycled coarse aggregate. (Chinese patent code: ZL 2021 1 0316969.X).

Author Contributions: L.Y.: Conceptualization, Data curation, Investigation, writing-original draft, Writing—review and editing. S.L.: Funding support. Z.Z.: Writing—review and editing. Z.L.: Funding acquisition. All authors have read and agreed to the published version of the manuscript.

Funding: This work has been supported by two grants. One is from A Special Fund Item of Scientific and Technological Innovation (Item number 2019-C534) which is from Research Institute of Highway Ministry of Transport, China. The second one is from Shandong Highway Incorporated Company to Research Institute of Highway Ministry of Transport, China. The Technology Plan Item (Item number 2018BZ3) set up by Transportation Department of Shandong Province, China.

Institutional Review Board Statement: Not applicable.

Informed Consent Statement: Not applicable.

Data Availability Statement: The data presented in this study are available on request from the corresponding author.

Acknowledgments: The raw materials were supported by Shandong Highway Incorporated Company to Research Institute of Highway Ministry of Transport, China.

Conflicts of Interest: The authors declare that they have no known competing financial interests or personal relationships that could have appeared to influence the word reported in this paper.

References

1. Quan, H.-Z. Application overview and technical standards of recycled concrete in other countries. *J. Qingdao Technol. Univ.* **2009**, *30*, 87–92.
2. Poon, C.S.; Kou, S.C.; Lam, L. Influence of recycled aggregate on slump and bleeding of fresh concrete. *Mater. Struct.* **2007**, *40*, 981–988. [CrossRef]
3. Jin, R.; Li, B.; Elamin, A.; Wang, S.; Tsioulou, O.; Wanatowski, D. Experimental Investigation of Properties of Concrete Containing Recycled Construction Wastes. *Int. J. Civ. Eng.* **2018**, *16*, 1621–1633. [CrossRef]
4. Leite, M.B.; do Filho, F.; Lima, P.R. Workability study of concrete made with recycled mortar aggregate. *Mater. Struct.* **2013**, *46*, 1765–1778. [CrossRef]
5. Revathi, P.; Selvi, R.S.; Velin, S.S. Investigations on Fresh and Hardened Properties of Recycled Aggregate Self Compacting Concrete. *J. Inst. Eng. India Ser. A* **2013**, *94*, 179–185. [CrossRef]
6. Ju, M.; Park, K.; Park, W.J. Mechanical behavior of recycled fine aggregate concrete with high slump property in normal-and high-strtength. *Int. J. Concr. Struct. Mater.* **2019**, *13*, 61. [CrossRef]
7. Yehia, S.; Helal, K.; Abusharkh, A.; Zaher, A.; Istaitiyeh, H. Strength and Durability Evaluation of Recycled Aggregate Concrete. *Int. J. Concr. Struct. Mater.* **2015**, *9*, 219–239. [CrossRef]
8. Mo, K.H.; Ling, T.-C.; Cheng, Q. Examining the Influence of Recycled Concrete Aggregate on the Hardened Properties of Self-compacting Concrete. *Waste Biomass-Valorization* **2021**, *12*, 1133–1141. [CrossRef]

9. Wang, Z.; Wang, L.; Cui, Z.; Zhou, M. Effect of recycled coarse aggregate on concrete com-pressive strength. *Trans. Tianjin Univ.* **2011**, *17*, 229–234. [CrossRef]
10. Sharma, R. Laboratory Study on Effect of Construction Wastes and Admixtures on Compressive Strength of Concrete. *Arab. J. Sci. Eng.* **2017**, *74*, 373–3962. [CrossRef]
11. Zheng, Y.; Zhang, Y.; Zhang, P. Methods for improving the durability of recycled aggregate concrete: A re-view. *J. Mater. Res. Technol.* **2021**, *15*, 6367–6386. [CrossRef]
12. Su, Y.; Gu, S.; Chang, X.-L.; Li, J.-X. Experiment study on recycled coarse aggregate modified with heating method. *China Concr. Cem. Prod.* **2015**, *10*, 6–10.
13. Xiao, J.Z.; Wu, L.; Fan, Y.H. Test on modification of recycled coarse aggregate by micro wave heating. *Concrete* **2012**, *7*, 55–57.
14. Chen, M.-Z.; Zhong, J.-J.; Wu, S.-P.; Zhu, J.-Q.; Wang, D.-M. Physical properties research of modified recycled coarse aggregate. *High Way* **2011**, *7*, 198–201.
15. Li, L.; Xuan, D.; Chu, S.H.; Poon, C.S. Modification of recycled aggregate by spraying colloidal nano silica and silica fume. *Mater. Struct.* **2021**, *54*, 223. [CrossRef]
16. He, D.-Z. Japanese Concrete strengthening treatment technology. *Spec. Struct.* **2000**, *17*, 39.
17. Zeng, W.; Zhao, Y.; Zheng, H.; Poon, C.S. Improvement in corrosion resistance of recycled aggregate concrete by nano silica suspension modification on recycled aggregates. *Cem. Concr. Compos.* **2020**, *106*, 103476. [CrossRef]
18. Zhu, Y.N.; Zhang, H.R.; Meng, T.; Zhao, Y.X. Performance of nano-SiO_2 modified recycled aggregate concrete applied in a real project. *Concrete* **2014**, *7*, 138–144.
19. Li, K.; Zhang, B.; Yang, D.; Xiao, L.; Xia, J.; Zhou, Y. Research on Influence of Silane Emulsion Impregnation on Properties of Concrete Made of African Natural High Water Absorption Aggregate. *Subgrade Eng.* **2016**, *5*, 114–118.
20. Li, Y. *Improvement of Recycled Concrete Aggregate Properties by CO_2 Curing*; Hunan University: Changsha, China, 2014; Volume 5.
21. Wang, J.H.; Geng, O.; Li, F.M. Experiment on effect of several measures by modifying recycled coarse ag-gregate on improving compressive strength of concrete. *J. Archit. Civ. Eng.* **2016**, *33*, 91–97.
22. Choi, H.; Choi, H.; Lim, M.; Inoue, M.; Kitagaki, R.; Noguchi, T. Evaluation of the mechanical perfaormance of low-quality recycled aggregate through interface enhancement between cement matrix and coarse aggregate by surface modification technology. *Int. J. Concr. Struct. Mater.* **2016**, *10*, 87–97. [CrossRef]
23. Shaban, W.M.; Yang, J.; Su, H.; Liu, Q.F.; Tsang, D.C.; Wang, L.; Xie, J.; Li, L. Properties of recycled concrete aggregates strengthened by different types of pozzolan slurry. *Constr. Build. Mater.* **2019**, *216*, 632–647. [CrossRef]
24. Wang, Y.; He, F.; Wang, J.; Hu, Q. Comparison of Effects of Sodium Bicarbonate and Sodium Carbonate on the Hydration and Properties of Portland Cement Paste. *Materials* **2019**, *12*, 1033. [CrossRef]
25. Wang, Y.; He, F.; Wang, J.; Wang, C.; Xiong, Z. Effects of calcium bicarbonate on the properties of ordinary Portland cement paste. *Constr. Build. Mater.* **2019**, *225*, 591–600. [CrossRef]
26. Wang, H.; Sun, X.; Wang, J.; Monteiro, P.J. Permeability of Concrete with Recycled Concrete Aggregate and Poz-zolanic Materials under Stress. *Materials* **2016**, *9*, 252. [CrossRef] [PubMed]
27. Xie, J.; Huang, L.; Guo, Y.; Li, Z.; Fang, C.; Li, L.; Wang, J. Experimental study on the compressive and flexural behaviour of recycled aggregate concrete modified with silica fume and fibres. *Constr. Build. Mater.* **2018**, *178*, 612–623. [CrossRef]
28. Shaban, W.M.; Yang, J.; Su, H.; Mo, K.H.; Li, L.; Xie, J. Quality Improvement Techniques for Recycled Concrete Aggregate: A review. *J. Adv. Concr. Technol.* **2019**, *17*, 151–167. [CrossRef]
29. Tam, V.W.; Gao, X.; Tam, C.M.; Chan, C. New approach in measuring water absorption of recycled aggregates. *Constr. Build. Mater.* **2008**, *22*, 364–369. [CrossRef]
30. Rajagopal, A. *Automated Laboratory Testing Methods for Specific Gravity and Absorption Verified to Match the Current Method Results Final Report*; Fine Aggregates; Infrastructure Management and Engineering Inc.: Cincinnati, OH, USA, 2009; p. 82.
31. Xing, J.; Yu, L. Selection of concrete pores filling materials and its applicability analysis. *New Build. Mater.* **2019**, *6*, 51–55.
32. Bhasya, V.; Bharatkumar, B.H. Mechanical and durability properties of concrete produced with treated recycled concrete aggregate. *ACI Mater. J.* **2018**, *115*, 209–217. [CrossRef]
33. *ASTM C127*; Standard Test Method for Density, Relative Density (Specific Gravity), and Absorption. ACM: New York, NY, USA, 2014.
34. *GB/T 14685-2011*; PRC National Standard. Pebble and Crushed Stone for Construction. SAC: San Bernardino, CA, USA, 2011.
35. *JGJ/T 322-2013*; PRC Building Industrial Standard. Technical Specification for Test of Chloride Ion Content in Concrete. China Building Industry Press: Beijing, China, 2013.
36. *ASTM C188*; Standard Test method for Density of Hydraulic Cement. ASTM: West Conshohocken, PA, USA, 2009.
37. *ASTM C204*; American Association State Highway and Transportation Officials Standard. ASTM: West Conshohocken, PA, USA, 2016.
38. Alibaba Website. Available online: https://detail.1688.com/offer/38086091044.html?spm=a2615.7691456.autotrace-offerGeneral.19.cdb64761jjVk5z (accessed on 16 February 2022).
39. Alibaba Website. Available online: https://detail.1688.com/offer/586664215978.html?spm=a312h.2018_new_sem.dh_002.7.fd22203fBgQA2H&cosite=baidujj_pz&tracelog=p4p&clickid=c56accd8033e4c69a22de5c1dc865c13&sessionid=1677a50ffad3af2e79ebedc3d04940e3 (accessed on 16 February 2022).
40. *GB/T 175*; PRC National Standard. Common Portland Cement. China Standards Press: Beijing, China, 2020.

41. *GB/T 14684;* PRC National Standard. Sand for Construction. China Standards Press: Beijing, China, 2011.
42. *ASTM C143/C143-M;* Slump of Hydraulic Cement Concrete. ASTM: West Conshohocken, PA, USA, 2012.
43. *ISO 4012:1978-11;* Concrete—Determination of Compressive Strength of Test Specimens. ISO: Geneva, Switzerland, 1978.
44. Xu, Q. The Research on Spatial and Frequency Distribution of the Pore Structure of Cement Based Materials. Master's Thesis, Southeast University of China, Nanjing, China, 2014.
45. Zhao, H.; Ding, J.; Huang, Y.; Tang, Y.; Xu, W.; Huang, D. Experimental analysis on the relationship between pore structure and capillary water absorption characteristics of cement-based materials. *Struct. Concr.* **2019**, *20*, 1750–1762. [CrossRef]

coatings

Article

Carbonation Resistance of Surface Protective Materials Modified with Hybrid NanoSiO_2

Kailun Xia [1], Yue Gu [1,*], Linhua Jiang [1], Mingzhi Guo [1], Lei Chen [1] and Feilong Hu [2]

[1] College of Mechanics and Materials, Hohai University, Nanjing 210098, China; xklhhu@hhu.edu.cn (K.X.); lhjiang@hhu.edu.cn (L.J.); mzguo@hhu.edu.cn (M.G.); chenyuanshi@hhu.edu.cn (L.C.)
[2] Suzhou Hechuan Chemical Technology Service Co., Ltd., Suzhou 215216, China; feihoo15@163.com
* Correspondence: gubetter@hhu.edu.cn

Abstract: To date, reinforcement concrete is the main construction material worldwide. As the concentration of atmospheric CO_2 is steadily increasing, carbonation of the reinforcement concrete becomes a pressing concern. In this study, novel surface protective materials (SPMs) modified with hybrid nanoSiO_2 (HNS), fly ash, and slag were developed to reduce CO_2 emissions and extend the service life of the reinforcement concrete. The carbonation depths were measured by phenolphthalein to reflect the carbonation resistance. X-ray diffraction (XRD), fourier-transform infrared spectroscopy (FTIR), and thermal gravimetric analysis (TGA) were conducted to analyze the chemical components of the samples after carbonation. In addition, MIP was carried out to examine the microstructures of the samples prior to carbonation. Thermodynamic modeling was employed to calculate the changes in the phase assemblages of each blends in an ideal situation. The experimental results showed that the carbonation depth and $CaCO_3$ content of the SPM modified with HNS decreased by 79.0% and 64.6% compared with the reference, respectively. The TGA results showed that after carbonation, the $CaCO_3$ contents were 4.40% and 12.42% in the HNS modified samples and reference samples, respectively. MIP analysis demonstrated that the incorporation of HNS in SPM led to a 48.3% and 58.5% decrease in big pores and capillary pores, respectively. Overall, the SPMs modified with HNS in this study possessed better carbonation resistance and refined pore structures.

Keywords: reinforcement concrete; carbonation; hybrid nanoparticles; pore structure

1. Introduction

Over the past decade, the concentration of CO_2 in the atmosphere has increased significantly [1], resulting in associated environmental pollution problems. As the binder of reinforcement concrete, cement has been widely used across the world. However, the production of cement generates a huge amount of anthropogenic CO_2 [2–4]. The cement industry as a whole accounts for about 5–8% of global CO_2 emissions. Thus, extending the service life of concrete is a viable strategy to reduce the demand of cement, which in turn will effectively cut the carbon footprint.

It is well-known that mature concrete has a tendency to react with the atmospheric CO_2 in natural environments. The carbonation of concrete is more prone to occurring upon exposure to a high concentrated CO_2 level, because CO_2 can readily diffuse into the pore solutions through a normally porous surface structure of concrete. The precipitation of $CaCO_3$ depletes Ca^{2+} in the pore solution, resulting in the release of Ca^{2+} from hydration products to maintain the dissolution equilibrium of calcium in the pore solution. This inevitably leads to changes in the hydration products. Among the hydration products, calcium hydroxide (CH) and calcium silicate hydrate (C–S–H) are the main reactants in the carbonation process (Equations (1) and (2)). For C–S–H, the depletion of Ca^{2+} tends to cause decalcification and decrease the Ca/Si ratio, forming decalcified C–S–H and eventually amorphous silica gel.

$$Ca(OH)_2 + CO_2 \rightarrow CaCO_3 + H_2O \tag{1}$$

$$C\text{-}S\text{-}H + 2CO_2 \rightarrow SiO_2 + 2CaCO_3 + H_2O \tag{2}$$

Another problem caused by the carbonation of concrete is the drop in the pH of the pore solution due to the neutralization reaction between carbonic acid and alkali. Under such a circumstance, the passive film on the steel surface has a propensity to undergo chloride-induced corrosion [5,6]. This poses a great threat to the durability of reinforcement concrete. Therefore, enhancing the carbonation resistance of reinforcement concrete represents an effective way to improve its durability and extend the service life.

In recent years, surface protective materials (SPMs) have attracted increasing attention as a feasible and economic way to prevent the diffusion of CO_2 into the inner parts of concrete, especially the vicinity of the reinforced steel [7,8]. On the other hand, supplementary cementitious materials (SCMs), such as fly ash and slag, are reported to be effective in augmenting the resistance of concrete to carbonation and chloride ingress [9]. At the same time, new cementitious binder materials composed of a high content of SCMs were referred to as lower carbon emission in cement manufacture [10]. Some researchers even found that some sorts of blast furnace slag can provide a better performance than the reference concrete [11]. This reasonably raises the question of whether the combination of SCMs with SPMs can synergically enhance the carbonation resistance of reinforcement concrete. However, one downside of the addition of SCMs in cementitious material is the negative effect on the early strength [12]. NanoSiO_2 (NS) has been extensively investigated to evaluate its ability to compensate for the SCMs-induced adverse influence on the early-age properties [13,14]. As an inorganic agent for modifying cement-based materials, NS holds great potential for commercial applications due to its low cost and excellent properties. For example, NS can elicit a filling effect because it can act as a micro aggregate to fill the aperture between the cement particles and thereby densify the microstructures [15,16]. Moreover, NS also has a seeding effect by adsorbing calcium ions and serving as nucleation sites of C–S–H due to the large specific surface area [17]. Furthermore, NS can function as a pozzolana because it can react with CH to form C–S–H [18,19].

The above three effects of NS work together to improve the early mechanical properties of reinforcement concrete, such as flexural strength and compressive strength [20,21]. As nano particles, however, NS is prone to agglomeration in the pore solution in the presence of various positive ions. In contrast, hybrid nanoSiO_2 (HNS) is a novel nanoparticle which has better dispersion than traditional NS. For example, Gu et al. [22] synthesized a core-shell nanoparticle and found that the obtained hybrid NS was well dispersed in water and remained stable. Collodetti et al. [23] found that the siloxane modified nanoSiO_2 densified the nanostructure of Portland cement pastes, and coating siloxane on the surface enhanced the nanoSiO_2 stability in pore solutions. Mora et al. [24] developed novel hybrid silica particles functionalized with n-dodecyl groups ($-C_{12}H_{25}$) and found that they both had better dispersion and hydrophobicity. Therefore, hybrid nanoSiO_2 with an organic-inorganic core shell structure seems to be an effective additive to promote microstructure and durability of cement-based materials. Since surface is the first barrier to resist carbonation, coating the surface with SPMs is a cost-effective way to enhance the carbonation resistance of concrete.

In this paper, surface protective materials (SPMs) were prepared with the incorporation of two types of SCMs, i.e., fly ash and slag. In order to compensate for the adverse effect of SCMs on the early-age properties, a kind of hybrid nanoSiO_2 (HNS) with a core-shell structure was also added to the SPM. The carbonation depth of the SPM was measured to assess its carbonation resistance ability. X-ray diffraction (XRD), thermal gravimetric analysis (TGA), and fourier-transform infrared spectroscopy (FTIR) were employed to characterize the changes in the chemical compositions of the SPM before and after carbonation. Moreover, mercury intrusion porosimeter (MIP) was conducted to investigate the changes in the pore structures before carbonation in different blends. Finally, thermodynamic modeling was used to simulate the changes in phase assemblage, pH value, and Ca/Si ratios upon interacting with CO_2.

2. Materials and Methods

2.1. Materials

The cement used in all the experiments was Ordinary Portland Cement (OPC) compliant with the Chinese Nation Standard GB8076-2008 which was purchased from Anhui Conch Cement Co., Ltd. (Wuhu, China). Table 1 lists the chemical compositions of OPC characterized by XRF. The class F fly ash and blast furnace slag were used. The chemical compositions of the used fly ash and slag are listed in Tables 2 and 3, respectively. The aggregate used was purchased from China ISO Standard Sand manufactured according to ISO679: 1989, EN196-1.

Table 1. Chemical compositions of Ordinary Portland Cement (OPC).

Component	CaO	SiO_2	Al_2O_3	Fe_2O_3	SO_3	MgO	K_2O	L.O.I.
Content(wt%)	62.83	20.50	5.61	3.84	3.07	1.70	1.31	1.14

Table 2. Chemical compositions of fly ash.

Component	CaO	SiO_2	Al_2O_3	Fe_2O_3	SO_3	MgO	K_2O	L.O.I.
Content(wt%)	3.39	57.23	28.34	4.07	1.08	1.22	2.51	1.02

Table 3. Chemical compositions of slag.

Component	CaO	SiO_2	Al_2O_3	Fe_2O_3	SO_3	MgO	K_2O	L.O.I.
Content(wt%)	26.51	46.29	7.84	5.05	0.62	10.46	1.76	1.47

The commercially available hybrid nanoSiO_2 (HNS) was provided by Jiangsu Sobute New Materials Co., Ltd. (Nanjing, China). The inorganic constituent of HNS is nanoSiO_2 (NS), while the organic components in HNS are mainly the aliphatic molecular group, which accounts for 13% (by weight), according to the instructions provided by the supplier. The diameter of HNS ranged from 30 to 60 nm. The schematic diagram of the molecular structure of HNS is presented in Figure 1.

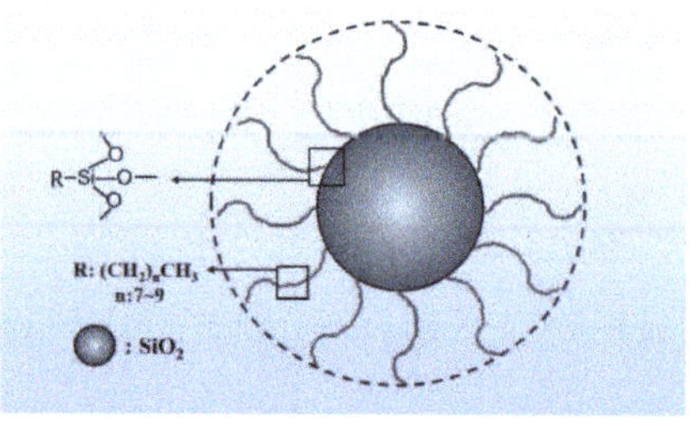

Figure 1. Schematic diagram of molecular structure of hybrid nanoSiO_2 (HNS).

2.2. Sample Preparation

Figure 2 shows the schematic diagram of applying the protective layer on the mortar substrate. The mortar substrate was first prepared by casting the well proportioned fresh mixture in a cubic mold with dimensions of 70 × 70 × 50 mm until initial setting. Then,

SPMs with a thickness of 20 mm were coated on the top surface of the substrate. All the samples were cured in a standard chamber (20 ± 2 °C, 95% relative humidity) for 28 days. Two types of SPMs were adopted in this experiment. The formula of both substrates and SPMs are listed in Table 4. The FA, slag, and HNS were added by replacing the same dosage of cement. All the abbreviations used in this study were listed in Table 5.

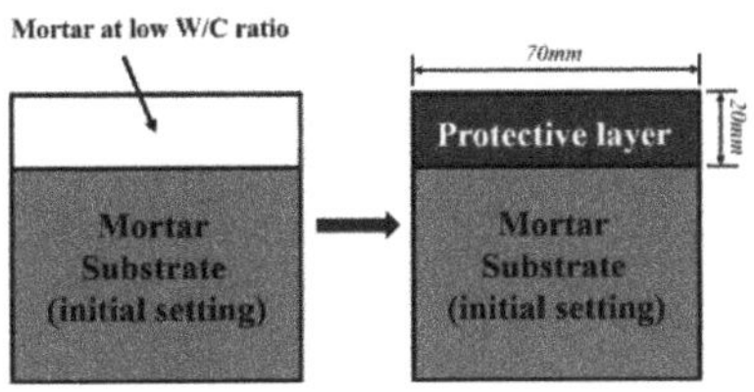

Figure 2. Schematic diagram of the process of applying the protective layer onto the mortar substrate.

Table 4. Mix proportions of substrate and different kinds of protective layers.

Parts	Water/Binder	Sand/Binder	FA(wt%)	Slag(wt%)	HNS(wt%)
SPM-0	0.40	3:1	15	15	0
SPM-1	0.40	3:1	15	15	1
Substrates	0.53	3:1	15	15	0

Table 5. Explanations of abbreviations.

Abbreviations	Content of Reference
SPMs	Srface protect materials
SCMs	Supplementary cementitious materials
NS	NanoSiO_2
HNS	Hybrid nanoSiO_2
FA	Fly ash
CH	$Ca(OH)_2$
XRD	X-ray diffraction
TGA	Thermal gravimetric analysis
FTIR	Fourier-transform infrared spectroscopy
MIP	Mercury intrusion porosimeter
REF	Samples consist of substrates without SPM
N	Samples consist of substrates and SPM-0
HN	Samples consist of substrates and SPM-1

HN represents the samples covered with protective layers consisting of 1 wt% HNS; N represents the samples covered with protective layers consisting of 0 wt% HNS; REF represents the samples without protective layers.

2.3. *Test Methods*

2.3.1. Carbonation Depth

Five sides of samples were sealed with paraffin wax (as shown in Figure 3), and the surface covered with the protective layer was exposed to CO_2 at a concentration of (20 ± 2)% and relative humidity of (70 ± 5)% in a HTX-12X carbonation experiment chest manufactured by Suzhou Donghua Testing Instrument Co., Ltd. (Suzhou, China). The carbonation test procedure was in compliance with Chinese National Standard GBJ82-85.

After 28 days of carbonation, the samples were split into two halves, and a phenolphthalein alcohol solution was sprayed onto the fractured surface to identify the carbonation depth.

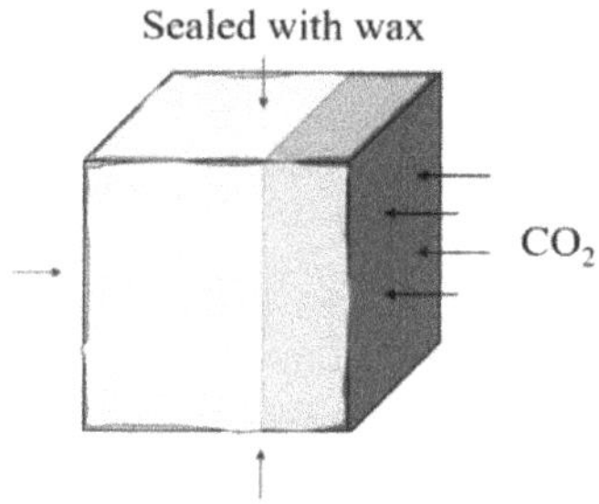

Figure 3. Direction of CO_2 diffusion during carbonation.

2.3.2. Chemical Composition

Specimens for chemical composition tests including XRD, TGA, and FTIR analysis were extracted from the SPMs within a 5 mm depth beneath the surface after carbonation and ceased hydration by immersing them into isopropanol for 48 h.

XRD was conducted to analyze the changes of mineral compositions of the SPMs before and after carbonation. A Bruker D8 Advance X-ray diffractometer (Bruker Technology Co., Ltd., Karlsruhe, Germany). was used, and the scanning range (2θ) was from 5° to 70° at a rate of 5°/min in a θ–θ configuration using Cu–Kα radiation.

TGA was carried out to investigate the mineral compositions of the SPMs before and after carbonation. Specimens for TGA were extracted from the carbonated area beneath the surface of SPMs and ground to powder followed by drying in 50 °C for 24 h. The weighted powder samples were tested in a nitrogen atmosphere by a NETZSCH STA449F3 thermal analyzer (NETZSCH-Gerätebau GmbH, Selb, Germany), and were heated from 30–1000 °C at a rate of 20 °C/min.

FTIR was carried out to analyze the changes of chemical compositions of SPMs after carbonation, a Nicolet IS10 fourier transform infrared spectrometer (Thermo Fisher Scientific, Waltham, MA, USA). was used to conduct the FRIR test. The wavenumbers range was 2000 to 400 cm^{-1} at a resolution of 2 cm^{-1}. Each sample was scanned 32 times.

2.3.3. Pore Structures

To investigate the pore structures of different SPMs before carbonation, a Micrometrics AutoPore IV 9510 (American Michael Instruments Corp., Charlottesville, VA, USA) was used to conduct the MIP test with a maximum pressure of up to 415 MPa. Specimens for the MIP test were extracted from the SPMs within a 5 mm depth beneath the surface before carbonation, and then stopped hydration by immersing them into isopropanol for 48 h. Pore size distribution curves ranging from 3~360 μm were drawn out.

2.3.4. Thermodynamic Modeling

A Gibbs free energy minimization program was used to calculate the changes of the phase assemblage during carbonation in a chemical system. CEMDATA18 database was used [25].

3. Results and Discussion

3.1. Carbonation Depth

Figure 4 shows the carbonation depth of different samples after 28 days of carbonation. It can be seen that REF has the largest carbonation depth of about 14.3 mm. Such a high carbonation degree may be due to a porous surface structure caused by a relatively large W/C ratio of 0.53. It is well-known that the W/C ratio is a vital parameter affecting the density of cementitious materials. Normally, a larger W/C ratio results in a looser microstructure and thereby provides more diffusion channels for CO_2 [26]. Compared with REF, the carbonation depth of group N (4.5 mm) and HN (3mm) is reduced by 69.2% and 79.0%, respectively. Apparently, the application of protective layers together with a low W/C ratio results in both a denser surface and a less amount of connected pores, impeding the diffusion of CO_2 and elevating the carbonation resistance. The highest carbonation resistance of HN may due to the pozzolanic reaction between HNS and $Ca(OH)_2$, leading to the formation of dense C–S–H and the consumption of $Ca(OH)_2$. This is because C–S–H is beneficial to densifying the microstructure of cement paste, while the presence of less $Ca(OH)_2$ decelerates the carbonation reaction. Moreover, the filling effect of nanoSiO_2 is also believed to be conducive to reducing the porosity.

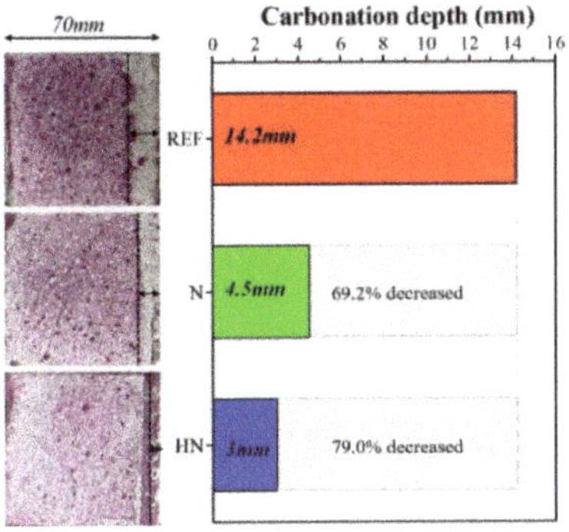

Figure 4. Carbonation depth of different samples after 28 days of carbonation.

From the results above, it can be concluded that the combination of a protective layer with a low W/C ratio can effectively reduce the carbonation of traditional cement-based materials. Interestingly, the use of HNS can further enhance this protective effect by the induced pozzolanic reaction and filling effect.

3.2. XRD

The XRD patterns of different samples after 28 days of carbonation are shown in Figure 5. The appearance of quartz peaks in Figure 5a is attributed to the presence of sand in the powdered samples, which is hard to be completely eliminated before the XRD test. Typical calcite peaks can be observed at 24.1°, 29.5°, 36.6°, and 39.5°. Obviously, REF has the highest content of calcite, while N has the lowest calcite content.

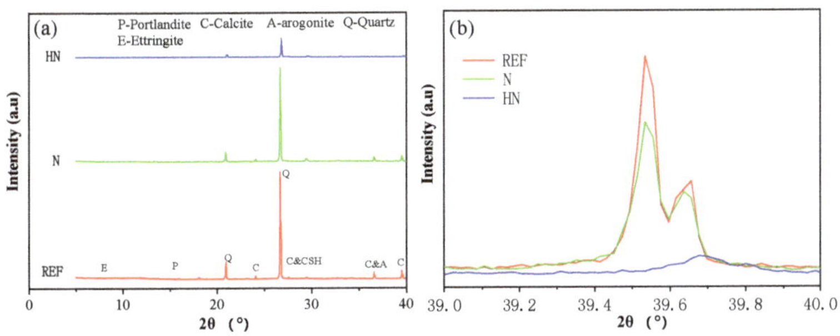

Figure 5. XRD patterns of different samples after carbonation: (**a**) overall patterns and (**b**) partial enlarged view of patterns near 39°.

Figure 5b shows the partial enlarged view of patterns near 39° with two remarkable calcite peaks. The calcite peaks at 39.5° and 39.6° reveal the highest carbonation degree of REF. Compared with REF, N experiences a drop in the calcite content. Moreover, the calcite peaks in HN are almost unobservable. A decrease of calcite in HN and N points to a better anti-carbonation performance compared with REF. The highest carbonation resistance of HN may be due to the previously mentioned improvement in the pore structures caused by the filling effect of HNS and the formation of dense C–S–H as a result of the pozzolanic reaction between HNS and CH. Plus, the consumption of CH by the pozzolanic reaction directly reduces the formation of calcite.

3.3. FTIR

The FTTR spectra patterns (ranging from 2000–400 cm^{-1}) of different samples are given in Figure 6. In each curve, three gray areas are used to highlight three bands: the band at 1440 cm^{-1} corresponds to $\upsilon3\text{-}CO_3^{2-}$, the band at 875 cm^{-1} corresponds to $\upsilon2\text{-}CO_3^{2-}$, and the band at 712 cm^{-1} corresponds to $\upsilon4\text{-}CO_3^{2-}$. The intensity of these three bands is in an increasing order of REF < HN < N. The band at 712 cm^{-1} is almost invisible in HN. A decrease in the band intensity reflects a decrease in the $CaCO_3$ content and a corresponding improvement in the carbonation resistance. The FTIR analysis demonstrates that HN has the best carbonation resistance.

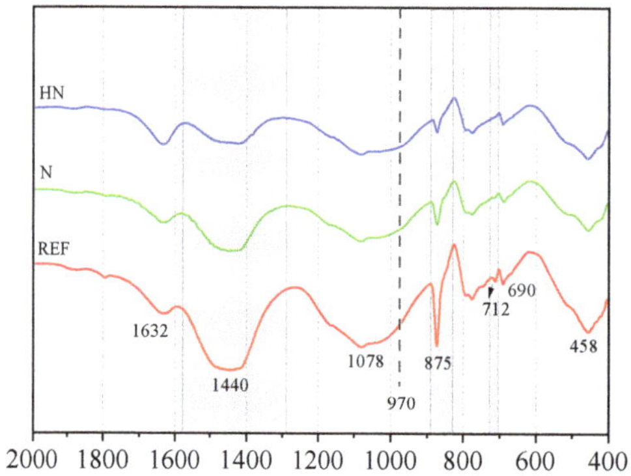

Figure 6. FTIR patterns of different samples after carbonation.

The dashed line in Figure 6 locates in a position at 970 cm^{-1}, which represents a band corresponding to the Si–O stretching vibration in C–S–H. At a Ca/Si of ≈1.7, the frequency of this band will increase along with the decalcification of C–S–H [27]. In Figure 6, the bands near 970 cm^{-1} in all the three patterns shift to a higher frequency area at 1078 cm^{-1}, while the position at 1080 cm^{-1} corresponds to amorphous silica gel, which is the final product of the decalcified reaction of C–S–H. This observation indicates the occurrence of different degrees of C–S–H decalcification: the stronger band at 1078 cm^{-1} reflects a larger content of amorphous silica gel and indicates a higher degree of decalcification. Taken together, the decalcification degree of different samples is in the decreasing order of HN < N < REF. The lower decalcification degree of HN could be attributed to the dense microstructure of C–S–H gels, further supporting the improved carbonation resistance induced by the incorporation of HNS.

The band at 1632 cm^{-1} is assigned to the bending vibration of H–O–H in chemical bound water in hydration products [28]. A decrease in the intensity of this band reflects the consumption of hydration products and a corresponding increase in the carbonation degree. The presence of the most remarkable band at 1632 cm^{-1} in Figure 6 once again confirms the best carbonation resistance of HN.

3.4. TG

The results of (Thermal Gravimetric Analysis) TGA of different samples carbonated for 28 days are shown in Figure 7. Mass loss peaks of different mineral compositions are observed in different temperature ranges: the peaks in the range of 100–380 °C correspond to the dehydration of C–S–H/Aft and the peaks within 380–460 °C correspond to the dihydroxylation of $Ca(OH)_2$ (Equation (3)); the peaks emerging from 460–750 °C correspond to the decomposition of $CaCO_3$ (Equation (4)).

$$CaCO_3 \rightarrow CaO + CO_2 \quad (3)$$

$$Ca(OH)_2 \rightarrow CaO + H_2O \quad (4)$$

$$m(CaCO_3) = \frac{M(CaCO_3)}{M(CO_2)} \times m(CO_2) \quad (5)$$

$$m(Ca(OH)_2) = \frac{M(Ca(OH)_2)}{M(H_2O)} \times m(H_2O) \quad (6)$$

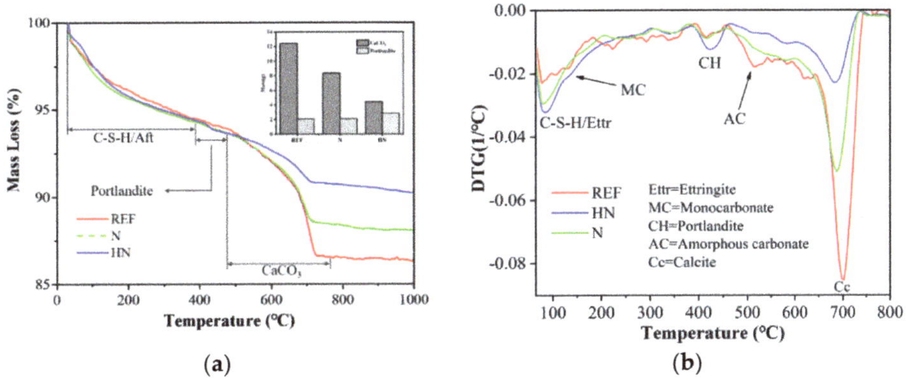

Figure 7. TG and DTG curves for different samples after carbonated for 28 days: (**a**) TG and (**b**) DTG.

Figure 7a shows the mass loss curves measured by TG, and the histogram in the legend presents the content of $Ca(OH)_2$ and $CaCO_3$ calculated by Equations (5) and (6), which is summarized in Table 6. The content of $CaCO_3$ in N and HN decreases by 32.60% and 64.47% relative to REF, respectively, indicating a decrease in the capacity of CO_2 absorption with the incorporation of HNS. This conclusion is further supported by the presence of a higher content of CH in HN compared with N and REF.

Table 6. Content of $CaCO_3$ and $Ca(OH)_2$ in different samples after carbonation.

Components	REF	N	HN
$CaCO_3$(wt%)	12.42	8.37	4.40
$Ca(OH)_2$(wt%)	2.05	2.11	2.85

As can be seen from Figure 7b, the peak of $CaCO_3$ in REF extends to the low temperature area, and a small peak appears at about 500 °C, which probably corresponds to amorphous carbonates due to incomplete crystallization under a high degree of carbonation [29]. The presence of amorphous carbonates may be attributed to the carbonation of hydration products, such as C–S–H and ettringite [30, 31]. These observations are consistent with the results of the FTIR analysis, demonstrating a lower degree of decalcification in the samples with a protective layer and the incorporation of HNS.

The observed lower carbonation degree of HN is possibly due to the following two reasons: (a) The filling effect of HNS improves the pore structure and reduces the amount of connected pores, effectively impeding the CO_2 diffusion; (b) The formation of secondary C–S–H induced by the seeding effect of HNS further densifies the microstructure of the protective layer and thereby enhances the resistance to CO_2 diffusion. As shown in Figure 8, the accumulation of C–S–H on the surface of cement particles can effectively reduce the contact between the un-hydrated cement and water, thus retarding the migration of Ca^{2+} [32,33]. Under such a circumstance, the concentration of Ca^{2+} might be further decreased due to the consumption by carbonation. Thus, the subsequent precipitation of $CaCO_3$ becomes more difficult.

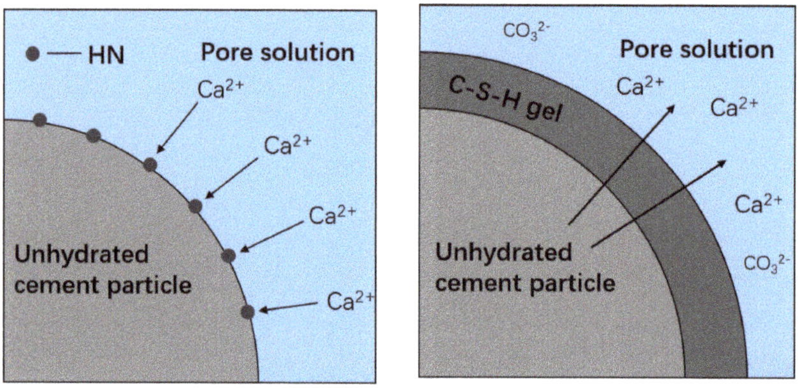

(**a**) (**b**)

Figure 8. Schematic diagram of Ca^{2+} migration: (**a**) during hydration and (**b**) during carbonation.

3.5. Pore Structures

Figure 9 shows the pore size distribution of different samples. Obviously, REF has the largest total porosity compared with N and HN due to its high W/C ratio. With the addition of HNS, HN has the lowest porosity. Pore volumes at different diameters for different samples are shown in Figure 10. Generally, the pores in concrete are classified into four categories: gel pores (≤10 nm), transition pores (10–100 nm), capillary pores (100–1000 nm), and big pores (>1000 nm) [34,35]. Big pores and capillary pores are harmful to the cement-based materials because they are the main diffusion channels for corrosive medium diffusion. It can be seen from Figure 10 that compared with REF, the volume of big pores in N and HN is decreased by 7.9% and 48.3%, respectively, while the volume of capillary pores in N and HN is, respectively, decreased by 38.2% and 58.5%. These results well demonstrate that the addition of HNS in cementitious materials not only reduces the total porosity but also refines the pore structures. The formation of secondary C-S-H induced by HNS fills in the voids between cement grains, resulting in the improvement of microstructures.

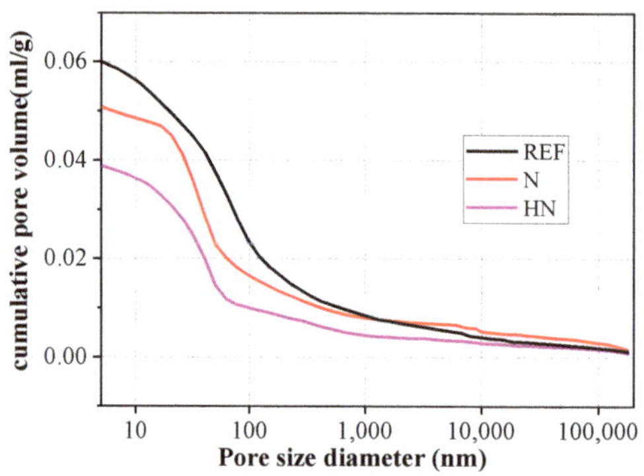

Figure 9. Pore size distribution of different samples.

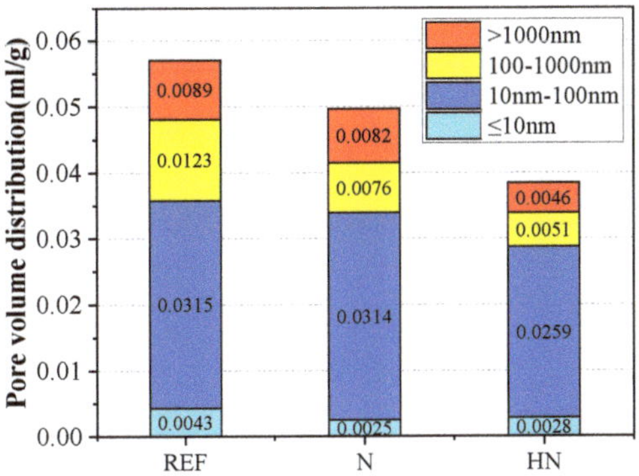

Figure 10. Pore volumes of different samples at different diameters.

3.6. Thermodynamic Modeling

Figure 11 shows the phase assemblage changes of two different blends in contact with CO_2 predicted by thermodynamic modeling. The ordinate represents the volume percentage of original solid phases as a function of the CO_2 content. As can be seen from Figure 11, N and HN display a similar graph, indicating that HNS has a negligible effect on the total volumes of each phase. It is worth noting that CH is not observed due to the SCMs induced pozzolanic reaction. This result is not consistent with previous studies. It is possible that the added SCMs are not completely reacted with cement in actual situations. Meanwhile, the stratlingite and high-Ca C–S–H become the dominating phases in a non-CO_2 situation. As the CO_2 content increases, stratlingite and C–S–H are consumed along with the formation of decalcified C–S–H, calcite and gibbsite.

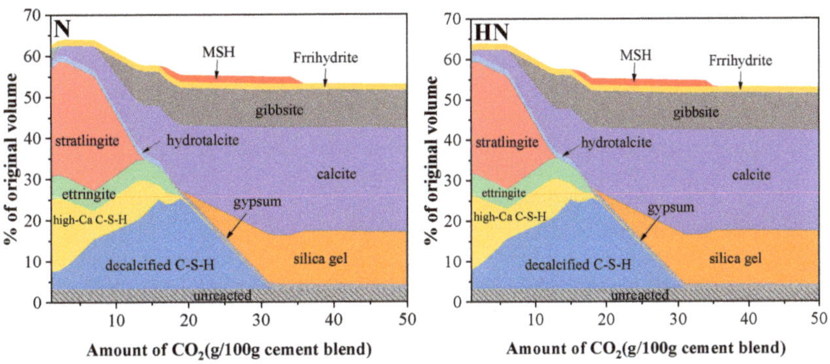

Figure 11. Changes of phase assemblage of N and HN samples.

Changes of pH and Ca/Si ratio of two blends in contact with CO_2 are shown in Figure 12. Two remarkable platforms are observed, which correspond to the consumption of stratlingite and decalcified C–S–H, respectively. This observation indicates that stratlingite and decalcified C–S–H can maintain a stable Ca/Si ratio and pH during carbonation. Thus, a high content of C–S–H and stratlingite may be beneficial to preventing the decrease of pH in cement-based materials upon contact with CO_2. This helps explain why the HNS incorporated samples possess better carbonation resistance in practical applications.

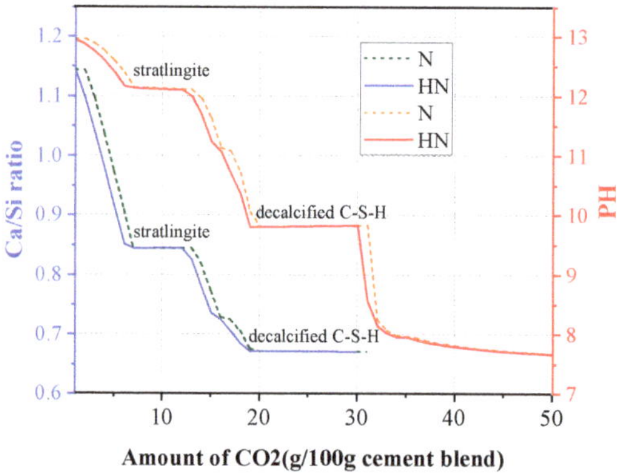

Figure 12. Changes of Ca/Si ratio and pH in contact with CO_2.

It can also be observed that the slope absolute values of both curves become smaller along with the decrease of the Ca/Si ratio, suggesting that a relatively smaller Ca/Si ratio is not conducive to decalcification. Therefore, the incorporation of HNS in cementitious materials is likely to retard the decalcification of C–S–H and stratlingite by increasing the Si/Ca ratio.

4. Conclusions

In this study, surface protective materials (SPM) with hybrid nanoSiO_2 (HNS) were prepared and applied on the surface of mortar substrates. The carbonation resistance of SPMs was investigated, and HNS was found to be effective in enhancing the carbonation resistance of cement mortar. The main conclusions can be drawn as follows:

1. The use of SPMs effectively increased the carbonation resistance of the substrates. Compared with REF, the carbonation depth of N and HN was reduced by 69.2% and 79.0%, respectively.
2. The combination of HNS with SPMs further enhanced the carbonation resistance. Compared with the SPMs without HNS, the carbonation depth of the HNS incorporated SPMs was reduced from 4.5 to 3 mm.
3. The incorporation of HNS in SPMs refined the pore structures. Compared with the SPMs without HNS, the HNS incorporated SPMs experienced a 43.9% and 32.9% reduction in big pores and capillary pores, respectively. This was ascribed to the formation of extra C–S–H and the filling effect of nanoparticles.
4. From thermodynamic modeling, it can be observed that the introduction of HNS could promote the formation of C–S–(A)–H, prevent the decrease of pH in pore solution, and lower the Ca/Si ratio of C–S–(A)–H. These effects are beneficial to augmenting the carbonation resistance of SPMs.

Overall, this study verified the enhancing effect of HNS on the carbonation resistance of cement mortar. However, the synergistic effect of HNS and SCMs, such as fly ash and slag, merits further research.

Author Contributions: Article writing, experimental data analysis, K.X.; methodology, Y.G.; conceptualization, L.J.; review and editing, M.G.; Software operation, L.C. processing experimental data, F.H. All authors have read and agreed to the published version of the manuscript.

Funding: This research was founded by the National Natural Science Foundation of China (Grant Nos.51808188).

Data Availability Statement: The data used to support the findings of this study are available from the corresponding author upon request.

Acknowledgments: The authors thank Jiangsu Research Institute of Building Science Co., Ltd.

Conflicts of Interest: The authors declare that they have no known competing financial interest or personal relationships that could appeared to influence the work reported in this paper.

References

1. Zhan, B.; Poon, C.; Shi, C. CO_2 curing for improving the properties of concrete blocks containing recycled aggregates. *Cem. Concr. Compos.* **2013**, *42*, 1–8. [CrossRef]
2. Thomas, J.J.; Biernacki, J.J.; Bullard, J.W.; Bishnoi, S.; Dolado, J.S.; Scherer, G.W.; Luttge, A. Modeling and simulation of cement hydration kinetics and microstructure development. *Cem. Concr. Res.* **2011**, *41*, 1257–1278. [CrossRef]
3. Xi, F.; Davis, S.J.; Ciais, P.; Crawford-Brown, D.; Guan, D.; Pade, C.; Shi, T.; Syddall, M.; Lv, J.; Ji, L.; et al. Substantial global carbon uptake by cement carbonation. *Nat. Geosci.* **2016**, *9*, 880–883. [CrossRef]
4. Andres, R.J.; Boden, T.A.; Bréon, F.M.; Ciais, P.; Davis, S.; Erickson, D.; Gregg, J.S.; Jacobson, A.; Marland, G.; Miller, J.; et al. A synthesis of carbon dioxide emissions from fossil-fuel combustion. *Biogeosciences* **2015**, *9*, 1845–1871. [CrossRef]
5. Richardson, M.G. *Fundamentals of Durable Reinforced Concrete*; Spon Press: New York, NY, USA, 2002.
6. Shah, V.; Scrivener, K.; Bhattacharjee, B.; Bishnoi, S. Changes in microstructure characteristics of cement paste on carbonation. *Cem. Concr. Res.* **2018**, *109*, 184–197. [CrossRef]
7. Seneviratne, A.M.G.; Sergi, G.; Page, C.L. Performance characteristics of surface coatings applied to concrete for control of reinforcement corrosion. *Constr. Build. Mater.* **2000**, *14*, 55–59. [CrossRef]
8. Ibrahim, M. Effectiveness of concrete surface treatment materials in reducing chloride-induced reinforcement corrosion. *Constr. Build. Mater.* **1997**, *11*, 443–451. [CrossRef]
9. Papadakis, V.G. Effect of supplementary cementing materials on concrete resistance against carbonation and chloride ingress. *Cem. Concr. Res.* **2000**, *30*, 291–299. [CrossRef]
10. Xue, L.; Zhang, Z.; Wang, H. Hydration mechanisms and durability of hybrid alkaline cements (HACs): A review. *Constr. Build. Mater.* **2021**, *266*, 121039. [CrossRef]
11. Parron-Rubio, M.E.; Perez-Garcia, F.; Gonzalez-Herrera, A.; Oliveira, M.J.; Rubio-Cintas, M.D. Slag substitution as a cementing material in concrete: Mechanical, physical and environmental properties. *Materials* **2019**, *12*, 2845. [CrossRef]
12. Lam, L.; Wong, Y.L.; Poon, C.S. Degree of hydration and gel/space ratio of high-volume fly ash/cement systems. *Constr. Build. Mater.* **2000**, *30*, 747–756. [CrossRef]
13. Yehdego, T.; Peethamparan, S. The role of nano silica in modifying the early age hydration kinetics of binders containing high volume fly ashes. In *Nanotechnology in Construction*; Springer International Publishing: Cham, Switzerland, 2015; pp. 399–405.
14. Gu, Y.; Ran, Q.; She, W.; Liu, J. Modifying Cement Hydration with NS@PCE Core-Shell Nanoparticles. *Adv. Mater. Sci. Eng.* **2017**, *2017*, 1–13. [CrossRef]
15. Shah, S.P.; Hou, P.; Konsta-Gdoutos, M.S. Nano-modification of cementitious material: Toward a stronger and durable concrete. *J. Sustain. Cem. Based Mater.* **2016**, *5*, 1–22. [CrossRef]
16. Du, H.; Du, S.; Liu, X. Durability performances of concrete with nano-silica. *Constr. Build. Mater.* **2014**, *73*, 705–712. [CrossRef]
17. Thomas, J.J.; Jennings, H.M.; Chen, J.J. Influence of nucleation seeding on the hydration mechanisms of tricalcium silicate and cement. *J. Phys. Chem. C* **2009**, *113*, 4327–4334. [CrossRef]
18. Sandrolini, F.; Franzoni, E.; Pigino, B. Ethyl silicate for surface treatment of concrete–Part I: Pozzolanic effect of ethyl silicate. *Cem. Concr. Compos.* **2012**, *34*, 306–312. [CrossRef]
19. Hou, P.; Qian, J.; Cheng, X.; Shah, S.P. Effects of the pozzolanic reactivity of nanoSiO_2 on cement-based materials. *Cem. Concr. Compos.* **2015**, *55*, 250–258. [CrossRef]
20. Hou, P.; Cheng, X.; Qian, J.; Zhang, R.; Cao, W.; Shah, S.P. Characteristics of surface-treatment of nano-SiO2 on the transport properties of hardened cement pastes with different water-to-cement ratios. *Cem. Concr. Compos.* **2015**, *55*, 26–33. [CrossRef]
21. Ghafari, E.; Costa, H.; Júlio, E.; Portugal, A.; Durães, L. The effect of nanosilica addition on flowability, strength and transport properties of ultra high performance concrete. *Mater. Eng.* **2014**, *59*, 1–9. [CrossRef]
22. Gu, Y.; Ran, Q.; Shu, X.; Yu, C.; Chang, H.; Liu, J. Synthesis of nanoSiO_2@PCE core-shell nanoparticles and its effect on cement hydration at early age. *Constr. Build. Mater.* **2016**, *114*, 673–680. [CrossRef]
23. Collodetti, G.; Gleize, P.J.P.; Monteiro, P.J.M. Exploring the potential of siloxane surface modified nano-SiO2 to improve the Portland cement pastes hydration properties. *Constr. Build. Mater.* **2014**, *54*, 99–105. [CrossRef]
24. Mora, E.; González, G.; Romero, P.; Castellón, E. Control of water absorption in concrete materials by modification with hybrid hydrophobic silica particles. *Constr. Build. Mater.* **2019**, *221*, 210–218. [CrossRef]
25. Lothenbach, B.; Kulik, D.A.; Matschei, T.; Balonis, M.; Baquerizo, L.; Dilnesa, B.; Miron, G.D.; Myers, R.J. Cemdata18: A chemical thermodynamic database for hydrated Portland cements and alkali-activated materials. *Cem. Concr. Res.* **2019**, *115*, 472–506. [CrossRef]
26. Siddique, S.; Naqi, A.; Jang, J.G. Influence of water to cement ratio on CO_2 uptake capacity of belite-rich cement upon exposure to carbonation curing. *Cem. Concr. Compos.* **2020**, *111*, 103616. [CrossRef]
27. Pan, X.; Shi, C.; Farzadnia, N.; Hu, X.; Zheng, J. Properties and microstructure of CO_2 surface treated cement mortars with subsequent lime-saturated water curing. *Cem. Concr. Compos.* **2019**, *99*, 89–99. [CrossRef]
28. Chen, T.; Gao, X. Effect of carbonation curing regime on strength and microstructure of Portland cement paste. *J. CO_2 Utilization* **2019**, *34*, 74–86. [CrossRef]
29. Šauman, Z. Carbonization of porous concrete and its main binding components. *Cem. Concr. Res.* **1971**, *1*, 645–662. [CrossRef]
30. Rostami, V.; Shao, Y.; Boyd, A.J.; He, Z. Microstructure of cement paste subject to early carbonation curing. *Cem. Concr. Res.* **2012**, *42*, 186–193. [CrossRef]

31. Shi, Z.; Lothenbach, B.; Geiker, M.R.; Kaufmann, J.; Leemann, A.; Ferreiro, S.; Skibsted, J. Experimental studies and thermodynamic modeling of the carbonation of Portland cement, metakaolin and limestone mortars. *Cem. Concr. Res.* **2016**, *88*, 60–72. [CrossRef]
32. Li, R.; Hou, P.; Xie, N.; Ye, Z.; Cheng, X.; Shah, S.P. Design of SiO_2/PMHS hybrid nanocomposite for surface treatment of cement-based materials. *Cem. Concr. Compos.* **2018**, *87*, 89–97. [CrossRef]
33. Powers, T.C. Structure and physical properties of hardened portland cement paste. *J. Am. Ceram. Soc.* **1958**, *41*, 1–6. [CrossRef]
34. Jennings, H.M.; Kumar, A.; Sant, G. Quantitative discrimination of the nano-pore-structure of cement paste during drying: New insights from water sorption isotherms. *Cem. Concr. Res.* **2015**, *76*, 27–36. [CrossRef]
35. Wang, Q.; Li, S.; Pan, S.; Cui, X.; Corr, D.J.; Shah, S.P. Effect of graphene oxide on the hydration and microstructure of fly ash-cement system. *Constr. Build. Mater.* **2019**, *198*, 106–119. [CrossRef]

MDPI AG
Grosspeteranlage 5
4052 Basel
Switzerland
Tel.: +41 61 683 77 34

Coatings Editorial Office
E-mail: coatings@mdpi.com
www.mdpi.com/journal/coatings

www.ingramcontent.com/pod-product-compliance
Lightning Source LLC
Chambersburg PA
CBHW062102220926
51850CB00010B/2856
* 9 7 8 3 7 2 5 8 6 7 3 2 5 *